FIRE LIGHTING THROUGH THE AGES – SCIENCE AND TECHNIQUE

ALESSANDRO CONTINI

Michael Terence
Publishing

This edition first published in paperback by
Michael Terence Publishing in 2020
www.mtp.agency

ISBN 9781913653323

CONTENTS

INTRODUCTION

To most anthropologists the discovery of fire marks the transition from ape to human because lighting and kindling a fire require skills that the earliest hominins did not possess. Fire has played a major role in the civilisation process of mankind. In prehistoric times it provided warmth and light, it enabled raw food to be cooked and gave protection from predators. In turn, a more digestible and nutritious cooked food diet sustained the gradual development of the human brain, causing it to expand from a volume of 600cc to 1.1L over approximately 500 thousand years. A larger brain was cognitively superior and capable of undertaking more complex tasks. During the iron and bronze ages fire was essential to smelt ores and cast spearheads and axe heads, which made hunting more efficient. After thousands of years we still use fire every day, although it is often disguised by the modern methods by which we produce it. We have bright light at the flick of a switch and instant warmth by regulating a thermostat. We cook food by turning a knob on the gas stove and we use fire to travel. The internal combustion engine of cars, ships and aircraft rely entirely on the energy produced by the confined burning of fossil fuel, while space missions are propelled by highly exothermic combustion of finely divided metals or hydrogen gas in the presence of strong oxidisers. Fire is literally part of our lives from cradle to grave, from lighting our first birthday-cake candle, to the incineration of our body at crematoria (as an alternative to burial).

Figure 1. Lighting candles in the Church of the Holy Sepulchre in the Old City of Jerusalem.

Going back a hundred years, steam engines and locomotives were coal fired as were most of the machinery used in factories during the industrial revolution that played an important role in the economic and social development of nineteenth-century Europe and the rest of the world. With the advent of alchemy, and eventually modern chemistry in the nineteenth century, mankind has gradually mastered the science of combustion phenomena, leaving behind inaccurate treaties like the phlogiston theory, which prevailed until the late eighteenth century. We have now tamed fire and even made it colourful. Evocative and elaborate firework displays of increasing beauty and magnitude represent only a small niche of pyrotechnics, the branch of science dealing with combustion reactions of a wide range of fuel and oxidiser systems.

Figure 2. An engraving from the 19th century showing factories and railways which relied on coal as fuel.

Fire is also used in another of mankind's objectionable activities, warfare. Delivering an explosive charge to the enemy requires a pyrotechnic sequence of reactions or 'fire train', that comprises an initiator coupled to a reliable propellant system. The first rudimentary rockets were invented by the Chinese well over a thousand year ago, following the allegedly accidental discovery of black powder by monks and alchemists.[i]

The idea to write this book came during one of the author's demonstrations of early fire lighting to friends and family. After having made fire by one of the primitive techniques described here, the same questions were inevitably asked: 'Blimey, how is that possible?' or 'How does it work?' and 'How did early human beings figure it out and when?'

There are publications dealing with early fire-lighting methods throughout history, but none describing the scientific principles behind each technique in simplified terms.[ii] This book aims to bridge that gap by describing early and modern fire-lighting methods while explaining the basic science behind each method. While useful references are given when required, the book avoids excessive jargon and maths in order to suit readers of all ages, from school children to fire-lighting and bushcraft enthusiasts. It is hoped that the book will also be useful to survival instructors, science teachers, experimental archaeologists and primitive technology experts. For the latter, there is an undeniable advantage in gaining an understanding of the science behind the techniques they may wish to experiment with. A useful analogy would be learning to drive while having prior knowledge of how the combustion engine works.

I hope the book will also inspire readers to safely experiment with some of the earliest lighting techniques described here, perhaps using foraged or improvised materials to re-enact historical fire making, or simulate a real survival scenario. Likewise, I encourage all readers to marvel at the long and fascinating history and technology of fire-lighting equipment through the centuries.

THE DISCOVERY OF FIRE

After 100 years of speculation, the archaeological consensus on the discovery of fire is still debated. Until a few decades ago, it was believed that Homo Erectus (standing human) first developed a technique to create fire at will more than 400,000 years ago. This would have followed an accidental discovery made earlier, possibly 1 million years ago. Fire remains, found at a very ancient site in Kenya, East Africa, suggest that our earliest ancestors had deliberately kindled fire in very remote times.[iii] The arrangement of stones surrounding fire remains, similar to those found at much later sites, are unequivocal signs of human-made fire.

Figure 3. Prehistoric humans gathering around a fire in a cave. (Artistic impression).

Mineralogical analysis of burnt clays found in caves indicate temperatures in excess of 900°C, which is a typical temperature reached by a camp fire, but normally too high for a fast sweeping bushfire. Similar research has been conducted in China at the Zhoukoudian cave, home of the famous Peking man, where siliceous aggregates (glass-like material) were found along with blackened and twisted bones showing signs of prolonged exposure to intense heat, consistent with being intentionally thrown into fire.[iv] Chemical analysis of the layers of blackened earth found immediately under the firing pit showed, indeed, traces of carbon, another strong indicator of combustion. To explain these very early fires, hypotheses have been suggested that layers of guano (accumulated bird droppings) could have ignited from spontaneous combustion (discussed later), but the very localised nature of the remains, coupled with traces of abundant flooding in the cave (which would have soaked combustible deposits), seem to rule out these unlikely theories.[v] So, if fire was already being produced in those ancient times, how was it obtained? Was it simply gathered from a smouldering tree stump which had been struck by lightning, or was it produced by wood-friction or stone-percussion methods? Sadly, nobody may find a definitive answer to these questions, as remains left behind from a fire started from natural sources would be identical to those of a fire started with an early wood-friction set.[vi] It is, however, possible to speculate that early humans might have discovered how to produce fire accidentally, perhaps after observing smoke produced when drilling a hole using a wooden spindle, or cutting grooves in logs using wooden tools. When working with stones, percussion of marcasite nodules with flint tools may have led to the observations of dull sparks. The most 'primitive' methods of fire lighting as they are generally known today, were probably developed by 'trial and error' attempts while carrying out other tasks.

Archaeological evidence about the use of fire by early humans is still being unearthed and debated at the time of writing. The recent discovery[vii] of large numbers of abraded nodules of manganese dioxide (MnO_2) scattered around archaeological Mousterian sites in France, have led to speculation that 50,000 years ago, some Neanderthals were using manganese dioxide dust to lower the ignition temperature of finely divided tinder, thus facilitating fire lighting by friction or percussion. In a series of controlled laboratory experiments, it was shown that manganese dioxide dust added to wood shavings lowers ignition temperature by more than 100°C. This is a consequence of the ability of MnO_2 to function as an oxidiser at relatively low temperature, thus transforming into Hausmannite, Mn_3O_4, which has indeed been recovered from the hearths of these Neanderthals caves. Infrared thermal imaging of the controlled combustion experiments also revealed that, while wood turnings do not ignite when heated up to 350°C, a mixture of turnings and manganese dioxide powder reliably ignites at temperatures around 250°C.

These findings demonstrate there is still much to learn about the use of fire by our ancestors.

NATURAL AND SPONTANEOUS FIRES

Fire, as a naturally occurring phenomenon, has made a regular appearance on earth since the planet had developed an atmosphere containing enough oxygen to sustain combustion reactions. After the great 'planet cool down' and formation of the oceans, blooming green algae colonies and other primeval photosynthesising organisms began to enrich the early atmosphere with oxygen gas, which was produced as a by-product of their metabolism. Initially, the primeval atmosphere would not have been able to sustain respiration on earth or any form of sustained burning, due to the relatively low (~5%) oxygen content. With the passage of millions of years, the composition of the earth's atmosphere became gradually more oxidising,[viii] with oxygen levels approaching the current value of 21% by volume, with the remainder being mostly nitrogen (78%) and small amounts of rare gases and carbon dioxide (~1%). Assuming a minimum oxygen concentration of 10% to support smouldering combustion, we can assume that fire, as a self-propagating reaction started by natural causes, would have been present on the planet long before the advent of the dinosaurs. So, what are the most common sources of ignition found in nature? These can be summarised into volcanic activity, cloud-to-ground lightning strikes and chemical self-heating processes ('spontaneous' combustion) that may occur inside large deposits of certain organic materials.

Volcanic Activity

In prehistoric times, volcanic activity would have been present only in certain areas of the globe, as is still the case today. Volcanos and magma-erupting vents are found both on land or under the sea along tectonic ridges between continental plates.

Figure 4. Mount Stromboli, southern Italy, during an eruption; igniting vegetation.

Magma (from the Greek 'thick substance') is essentially siliceous rock in liquid or semi-molten form containing dissolved gases, which is formed deep in the earth's mantle at temperatures in excess of 1,200°C. Magma is formed hundreds of kilometres underground due to the very high pressure and temperature of the mantle, which are controlled by the geothermal gradient and radioactive decay within the rock bed. Magma that leaves the underground and starts flowing on the surface is called 'lava'.

Figure 5. Mount Etna, southern Italy during an eruption, with lava flow.

Typical lava temperatures are well above the ignition point of dry vegetation, including solid wood. Direct contact with glowing lava isn't required for ignition, as radiant heat will easily dry and ignite vegetation at distances up to several metres. Another type of rare volcanic activity which may have been exploited as a natural ignition source, are self-combusting sulphur flows, which have recently been observed and documented at Vulcano island, in southern Italy.[ix] They consist of relatively long (a few hundred metres) rivulets of burning molten sulphur, rushing down the sloping fumarole fields of the volcano. Since the ignition temperature of sulphur in air at standard pressure is approximately 250°C, depending on purity, the streams of super-heated liquid sulphur ignite spontaneously and burn for a few hours with a faint blue flame, producing toxic sulphur dioxide gas.

Early humans would have had limited access to lava or burning sulphur as reliable fire-lighting tools. Only lava flows reaching the base of an active volcano would have been usable in this respect, as sulphur deposits are typically found closer to the crater, where fumarolic activity releases toxic gases which early humans would have avoided. Lightning strikes and spontaneous combustion would have been far more reliable and widespread natural sources of ignition in ancient times, especially in the Euro-Asiatic tundra and African planes, where early humans lived and foraged.

Lightning

Electrical storms have always ravaged the planet since the formation of a dense atmosphere gave rise to weather systems whose energy is fuelled by the sun heating the planet's surface. The physics of lightning strikes, which has recently been observed on other planets of the solar system, is well understood and documented.[x] Lightning is essentially a powerful electrical discharge occurring between two clouds or between the earth and a cloud.

Figure 6. An intense, forked cloud-to-ground discharge. Even the weaker branches are capable of setting objects alight.

The incredibly high voltage of the average lightning bolt (hundreds of millions of Volts) is typically discharged within microseconds through a highly luminous ionised plasma channel which reaches temperatures in excess of 20,000°C at its core. The diameter of the lightning channel varies widely between a few 0.5–2.5 cm, depending on the intensity of the transiting current, the duration of the discharge and the number of after-strokes. The formation of a typical single-stroke lightning bolt evolves through the birth of so-called 'stepped leaders', which are almost invisible, weakly-luminous ionised plasma paths or channels developing in the proximity of, or directly on top of protruding objects like rocks, bushes and trees.

Because of the electrical resistance of wood and the high temperature of the discharge channel, even the weakest of lightning bolts can potentially start a blaze. The Internet abounds with interesting amateur videos showing footage of Australian and Californian eucalyptus and palm trees burning fiercely after being hit by lightning. Footage of direct hits to human-made structures and vehicles offer a healthy reminder of the power of Mother Nature but also the risk of taking shelter under a tree during a thunderstorm. In all probability, early humans would have feared natural fires started by lightning strikes. With the passage of thousands of years, though, they eventually learnt to exploit natural ignition sources, thus taming fire. Moreover, a sweeping bush fire may leave behind partially cooked rodents and other small creatures that may have tasted better and were more digestible than raw meat. This is the probable link that brought early humans closer to natural fire than any other creature on the planet. That moment of great courage and determination to get closer to fire marked the start of an irreversible evolutionary journey, which set humanity firmly ahead of all other primates.

Self-Heating and Spontaneous Combustion

'Spontaneous' combustion originating from piled vegetable or carbonaceous matter is frequently observed in haystacks and barns containing haystacks.

Figure 7. Barn fire, possibly started by spontaneous combustion.

If a large enough (critical) mass of moist hay is piled up, it may start to self-heat from the core within a few days. Initial biological activity is caused by fermentation bacteria that are responsible for the production of heat within the pile. Heat accumulates due to the thermal insulating properties of the stack. Once the core temperature exceeds around 130°C, residual moisture is evaporated off and the bacteria are killed. From this point on, chemical oxidation takes over, aided by the gradual diffusion of air through the fibres of the stack. If the rate of heat evolution is greater than heat dissipated, temperatures in the core of the stack may reach in excess of 300°C, at which point smouldering auto-ignition may well occur. The stack starts burning from the core outwards and when the smouldering front reaches the outermost layers, where more air is naturally available, the whole pile may suddenly burst into flame. Every year, several barns and farm buildings are lost to fire due to the spontaneous combustion of hay stacks. For this reason, every farmer knows the danger of harvesting and piling large amounts of wet or moist hay. Moisture promotes bacterial proliferation within the pile, thereby starting the self-heating process.

A much faster oxidation reaction, which also leads to spontaneous combustion, is caused by the propensity of some vegetable oils to oxidise in the presence of air, releasing much heat in the process. Vegetable oils (and also animal fats) are formed by complex mixtures of triglycerides, which are triesters of glycerine (propane-1,2,3-triol) with long-chain 'fatty' carboxylic acids. The triglycerides found in some vegetable oils contain a very high percentage of unsaturated fatty acids, with up to three double carbon-to-carbon bonds in the chain. The higher the proportion of unsaturation of the oil (as measured by the 'Iodine number'), the greater its propensity for aerial oxidation and hence self-ignition under the right conditions. Highly unsaturated oils are known as 'siccative' oils (from the Latin, seccare, to dry), because they tend to polymerise during oxidation, forming a hard plastic-like substance. This property is precisely exploited when linseed oil is used to polish and protect furniture. However, a cotton rag soaked with linseed oil may in time spontaneously heat up and eventually ignite by thermal runaway, due to the same process which forms a durable sheen on freshly polished furniture.

The heat released by the oxidation reaction quickly builds up in the stained rag, especially if this is stored folded or crumpled up in a ball. The higher temperature, in turn, speeds up the rate of reaction, thus generating more heat and faster. When the rate of heat accumulation in the rag reaches the ignition point of the fabric, smouldering may begin. Every year, many house fires around the world are started from discarded oil-stained rags. The process is well documented in the literature.[xi] Typical fast self-heating oils include linseed, tung, cod liver and rapeseed but others are known to be moderately self-heating, they just take a little longer. Apart from liquid oils impregnated onto fabric, piles of mineral coal, peat and sometimes even oil seeds and animal manure are observed to self-heat during periods of hot weather and eventually, spontaneously ignite. Some underground lignite coal seams and peat bogs can burn for decades after self-ignition. The coal seam 'Burning Mountain' in New South Wales, Australia, is thought to have been on fire for 6,000 years after perhaps being ignited by lightning, and now provides an unusual attraction for tourists who flock to the area to observe the smoke billowing out of the ground.

The author has recently carried out a series of experiments to determine the minimum quantity of raw linseed oil deposited on a small cotton rag that could potentially ignite by self-heating. A single cotton rag (30x30 cm) smeared with only a few millilitres of raw linseed oil can amazingly self-combust within an hour, given the right conditions. Figure 8 illustrates how dangerous it can be to discard oil-stained rags in a dustbin, especially inside the home.

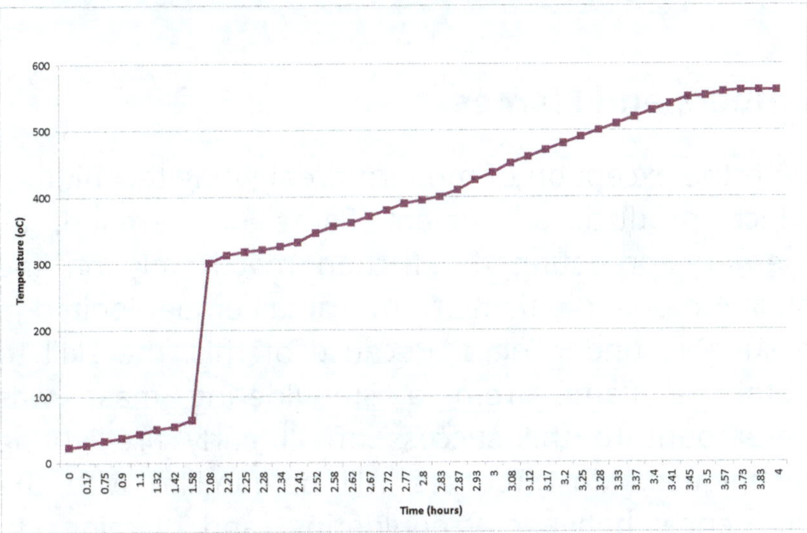

Figure 8. Temperature-monitoring of spontaneous ignition of a linseed oil-soiled rag, with temperature vs time profile. Note how the temperature suddenly jumped from 70°C to 300°C in less than 10 min. Transition from smouldering to flaming ignition occurred at 550°C after only 3.5h. No wonder that bins can catch fire during the night!

Despite this modern life hazard, would spontaneous combustion be a reliable ignition source during prehistoric times? Early humans may have occasionally come across fires started by self-heating processes. Habitation of caves where bats also nested, may have resulted in large piles of guano (bird droppings containing nitrates) spontaneously self-heating and eventually igniting, as argued by some archaeologists. However, haystacks, manure piles and vegetable oil-seed fires may not have been observed until at least the Bronze Age, with the advent of farming and agricultural activities. As such, self-heating as a reliable source of natural ignition may not be credible from an archaeological prospective.

Embers and Flames

With the exception of modern fire-lighting techniques, which produce an instant flame on demand, all methods predating the friction match rely on the ability to obtain a flame from a small ember ignited in a suitable tinder 'nest'. Because of this, the skill to obtain a flame from a smouldering mass was paramount to the success of all early fire-lighting techniques. Most readers will be familiar with the difference between smouldering and flaming. In general, the transition to flaming is marked by the sudden appearance of a luminous flame directly above a glowing ember, which, before the onset of flaming, typically produces smoke, a complex aerosol of pyrolysed fuel vapour, combustion gases and water vapour. A lit cigarette is a good example of a self-sustaining, flameless smouldering combustion front. No matter how hard the smoker inhales, the appearance of flame is highly unlikely on a lit cigarette, due to the high moisture content of fresh tobacco and the direction of travel (inside the cigarette) of the hot gases during the inhaling phase.

Figure 9. Lit cigarette.

Smouldering combustion processes can be modelled by complex mass and energy conservation equations which consider many parameters including diffused oxygen concentration, fuel and substrate properties (specific heat, density, heat conductivity, surface area, pore diameter) and many others. Speculation about the likely triggering mechanism to flaming include several gaseous reactions, including the so-called 'char oxidation' reaction and others, however, none has been conclusively proven to predominate.[xii] It is likely that most, if not all, mechanisms that depend on the nature of the fuel as well as atmospheric moisture, play a part in determining the onset of flaming from a smouldering mass.

Although we are all familiar with fire from everyday life, what exactly is a flame? A flame (from Latin, fiamma) is the light-producing gaseous part of a fire and is defined as a discrete region of space above a source of vaporised fuel where gas-phase oxidation reactions release heat and visible light (at least in most cases – hydrogen gas burns with an almost invisible flame), and sometimes sound (especially turbulent flames). The colour and temperature of a non-pyrotechnic flame (i.e. unlike those typically produced by fireworks) depend on the nature of the fuel being burnt and the extent of mixing of fuel vapour with air molecules. Two distinct types of flames are observed: premixed and diffusion flames.

*Figure 10. The appearance of premixed (blue)
and diffusion (yellow) flames.*

Premixed flames are normally human-made and formed when the air is mixed with fuel vapour in the correct ratio. These flames burn hotter and usually give off a blue colour, which is an indication of complete 'lean' combustion processes. A good example is the gas flame of the Bunsen burner with its air inlet fully open or a blow torch. Diffusion flames prevail in nature, and are formed by air molecules diffusing into the flame directly from its surrounding. As a consequence of the fuel and oxidant mixing mode, the oxidation reaction is not 'stoichiometric', i.e. not complete, and small soot particles always form. These particles incandesce, releasing yellow to orange visible light. A good example of a small diffusion flame is a candle flame. As the size of a diffusion flame increases, convection of the hot gases, less dense than the surrounding air, forces the flame to move from a laminar, i.e. ordered, to a more buoyant, turbulent burning regime. A large bonfire is a good example of this effect. The appearance of large diffusion flames is more turbulent and mobile as the size of the fire increases.

So when do embers transition to flaming?

Figure 11. A small, smouldering ember obtained by the author with a bow-drill friction fire set.

The first observation of a luminous gas-phase reaction zone is the necessary prerequisite condition to confirm the onset of flaming. Despite today's knowledge of fire chemistry, the transition from smouldering to flaming is an area which is still being actively investigated, mostly to limit the propensity of upholstered furniture to catch from lit cigarettes accidently dropped by careless smokers. The addition of flame retardants (usually organic compounds containing phosphorus and chlorine) to polyurethane-based foam fillers, delays and inhibits the spread of a smouldering front and therefore the onset of active flaming. This in turn gives more time for the occupants of a house to evacuate or try to fight the fire if it hasn't spread. An apparatus called 'cone calorimeter' is routinely used by fire chemists to measure the time to spontaneous or piloted flaming ignition of samples of furnishing materials when exposed to different radiant heat flux levels. Fire chemistry is a fascinating and life-saving research endeavour which should receive significantly more funding from all governments worldwide.

From the practical perspective involving most early fire-lighting techniques, a small smouldering tinder bundle always necessitates a forced supply of air though it to make the transition to flaming, whereas large smouldering fires are generally prone to flare up of their own accord, as they draw in enough air by convection and side suction. Forcing air into a dull-red glowing tinder bundle can be accomplished by carefully blowing air into it, or by moving the smouldering mass back and forward through the air, thus creating a relative, artificial breeze. Both techniques are equally effective. With more air molecules diffusing into the burning volume per unit time, the rate of the solid-phase reaction (i.e. smouldering) increases exponentially and generates much more heat per unit time, resulting in higher radiant heat flux, which is the reason an ember glows brighter and hotter when it's blown upon. The increased rate of smoke generation from a hotter ember (after being blown upon), is due to more volatile pyrolysis species and water vapour being released. The deadly gas carbon monoxide (CO) and other highly toxic compounds (hydrogen cyanide) are also given off (as in cigarette embers), therefore it is good practice to avoid inhaling fumes produced by smouldering fires, including cigarettes! Eventually, continuing to supply forced air into the smouldering mass, enough heat is accumulated to ignite the flammable pyrolysis vapours typically given off as acrid, white smoke. A small (10 cm^3) ember of dry grass and wooden twigs must glow bright orange to yellow (with temperatures exceeding 750°C, according to measurements) in order to flare up. Once a flame forms, the production of smoke normally ceases or is greatly reduced, as the fumes are consumed in the flame to form more benign carbon dioxide and water vapour. Depending on the size of the ember, transition to flaming may also be accompanied by an audible and muffled 'whoomph' sound, which is due to a sound wave, i.e. a front of compressed air molecules due to the sudden formation of the flame envelope. As the flame ignites, surrounding air molecules are effectively 'pushed out' of the way, thus forming a sound wave.

Lighting a fire from any smouldering tinder bundle takes lots of practice and requires good eye-to-arm coordination and judgement to know how long to continue blowing air over it, and how to build a hot ember in the first place. Excessively dense bundles, that have been compacted too hard, struggle to ignite because not enough air can be entrained. On the contrary, a loosely packed bundle may dissipate too much heat and will hence fail to ignite. Figure 12 shows photographs of a natural tinder bundle at various stages of the ignition process from a friction lighting technique.

Figure 12. Small jute tinder bundle undergoing transition from smouldering to flaming combustion.

CLASSIFICATION of FIRE-MAKING DEVICES THROUGHOUT HISTORY

Fire-making devices of all periods can be broadly classified into three main categories: mechanical (including friction and percussion methods), pyrochemical (including pyrophoric methods) and more recently, electrical (including resistive heating, piezo-electric and arc ignition).[xiii] In a few cases a particular method may be a hybrid between two categories, as we shall see later.

Early fire-lighting techniques were still somewhat 'primitive' even in the eighteenth century. Until the invention of the friction match, fire was still obtained mainly by mechanical (percussive) means. In more developed countries this was accomplished by the percussion of flint and steel, whereas primarily by wood friction in less advanced countries, particularly those blessed with a hot and dry climate. Although mainly done for ritual purposes, the Aboriginal Australians still produce fire by wood friction today, re-enacting the art of fire making as it was done by their ancestors. Several equally effective variants of the wood-friction method were in use. The choice of a particular method was often dictated by the type of wood locally available, while practical fire-lighting knowledge was passed down from one generation to the next, within families of nomadic groups.[xiv]

Less known but equally effective methods of fire lighting include solar (optical) ignition, used only during good weather by focussing the sun's rays with a burning lens or a concave metallic surface, and pneumatic (air) compression using wooden 'fire pistons' or 'fire syringes'. These techniques do not fall within friction or percussion categories and as such must be considered separately. Fire lighting by air compression is arguably one of the most fascinating and fastest methods of all the semi-primitive techniques. With a single downward stroke of a fire piston, the tinder is ignited at once, the whole process requiring almost no effort on the operator's part and very little energy expenditure.

Despite the existence of a number of recorded techniques, the archaeological literature suggests the earliest methods in use during the Palaeolithic and Mesolithic Ages involved percussion of stone on stone and wood friction. Both techniques remained in use in various parts of the globe for tens of thousands of years before the advent of the Iron Age. Both methods produce very small, fragile embers that must be skilfully coaxed into flame using a variety of foraged materials that easily smoulder, thus acting as ember 'extenders' (i.e. they easily extend the size of a small ember). Once enough heat had accumulated in the ember, suitable tinder like dried moss, finely shredded bark, dead leaves, dry grass and pollens, or even the shredded inner layers of several species of bracket fungi (which grow on the bark of certain species of sickly or dead trees), would be used as extenders to blow the initial small ember into flame. When dried, some species of bracket fungi can be easily set to smoulder and wrapped in birch bark to be blown into flame with remarkable ease.

With the advent of iron ore smelting, the predominant fire-lighting method was still based on mechanical percussion and involved striking flint or chert against a handheld forged steel tool. The technique is still commonly referred to as 'flint and steel'. Sparks generated by an iron striker are hotter than those given off by marcasite nodules, the carbon content of the steel being partly responsible (see later). The efficacy of this technique to ignite various types of tinder ensured that flint and steel remained in use in large parts of the globe until the advent of the friction match. During this long period throughout history, a great variety of shapes and designs of the steel striker were developed in different regions of the world, from the very basic Roman 'C' handle types to the very elaborate and ornate eighteenth-century 'Chuck Mucks' of Tibet, which incorporated a leather pouch to hold flint and tinder, thus providing a convenient and portable fire-lighting set. Throughout the early to late medieval period, almost every modest dwelling in Europe would have contained at least a flint and steel set, which was most likely kept in a wooden or (later) metallic tinder box on the fireplace hearth, to keep the tinder dry. Despite this, the pains and tribulations of some seventeenth-century maids to light an early morning fire using damp 'char cloth' tinder are well recorded. In fact the universal tinder of choice for flint and steel was charred rag or cloth, made from linen or cotton-based fabric carbonised over a fire in the absence of air to produce a high surface area carbon matrix which easily ignites from brief contact with incandescent metallic particles (sparks -see later). During the sixteenth century the use of steel strikers eventually culminated in the development of firearms (pistols and muskets) which relied on a 'flintlock' mechanism to ignite black powder (an intimate mixture of saltpetre, light-wood charcoal and sulphur). Similar to a manual strike-a-light set, the matchlock mechanism utilised a spring-loaded mechanism to strike a sharp flint edge against a steel surface. The resulting shower of sparks was directed into a small amount of powder held in the 'pan' of the gun. Following inflammation, fire was communicated to the propellant charge through a small hole drilled into the barrel of the weapon in proximity to the pan. Previous to this, the more unreliable and weather-sensitive matchlock ignition was used, which relied on a smouldering piece of rope that had been treated with dilute saltpetre (potassium nitrate), and placed in direct contact with the powder when firing. At the time, when flint lock firearms were in use, flint knapping became an essential industry, mostly artisanal in nature, as every fire arm required a regularly-shaped, sharp flint to operate correctly.

Replacement flints could be purchased from firearms dealers and army supply stores up until the advent of the percussion cap ignition device (early eighteenth century). But from this it can be seen how early fire-lighting techniques, the gunpowder industry and early military pyrotechnics were intimately connected. Excellent reviews of the gunpowder industry and flintlock firearms have been written, however, these topics are beyond the scope of this book. [xv]

Following the early Chinese discovery of black powder and other incendiary mixtures, the advent of alchemy and then early chemistry significantly advanced human knowledge of combustion reactions and thus fire-lighting techniques. Transition from the eighteenth-century phlogiston theory to the realisation of the role of atmospheric oxygen in the 'fire triangle', culminated in the discovery of flammable elements such as hydrogen, phosphorus and the development of new solid oxidisers to support combustion. The discovery of potassium chlorate in the eighteenth century eventually lead to the development of new chemical ignition methods of which the friction match was undoubtedly the most successful. Because of their simple and cheap manufacturing process, matches are still in use today in every country of the world, despite the wide diffusion of the modern disposal butane gas lighter. Liquid fuel (petrol) cigarette lighters started to rival matches as a means of producing fire only during the 1920s, after the introduction of the ferrocerium alloy (synthetic 'flint') and when availability of distilled hydrocarbon fuel became widespread. In the 1950s petrol lighters were largely replaced by mass-produced plastic ones containing liquefied butane gas. Today's cheap disposable plastic lighter symbolises the culmination of several thousand years of development in fire-lighting techniques, and the journey is far from over. Lighters that rely on a synthetic flint are to be categorised as mechanical percussion devices because of the need of a 'sparking wheel'. However, newer ignition methods relying on piezo-electricity and miniature tesla coil arc ignition, as well as optical methods involving diode lasers, are evolving at the time of writing. The following sections describe each technique in detail, explaining the scientific principles at their basis in a simplified manner.

Stone-on-Stone Percussion

Archaeological communities are divided when it comes to deciding whether early humans first obtained fire by means of wood friction or stone percussion. Wood friction, with its many variant methods, shows wide geographical distribution on all five continents, which leads some archaeologists to infer that it was the most common method of making fire in the remote past. Although the debate is beyond the scope of this book, the author has personally exchanged correspondence with Jurgen Weiner, a German experimental archaeologist who convincingly argues,[xvi] [xvii] along with others,[xviii] that the oldest proven method of intentional fire lighting in Europe was stone-on-stone percussion. According to Weiner, this was most likely done using 'sulphuric iron' (marcasite, a type of pyrite) which was struck with flint. Sparks would be 'caught' in tinder fungus (most likely *Fomes Fomentarius*). The oldest archaeological example of a Stone Age strike-a-light set was found in the Vogelherd cave in modern Germany and dates to about 30,000 BC. However, it is possible that an older example dating to 48000 BC, which was found in another cave located in modern Switzerland, may have been used for the same purpose. Evidence seems to conclusively prove that these pyrite nodules were indeed used for making fire, suggesting that European Neanderthals relied on this method on a daily basis. *Fomes Fomentarius*, the tinder fungus that would have likely been used to catch the feeble sparks generated by a Neanderthal's stone lighter, can be found in large areas of Eurasia, growing on birch (Betula) and beech (Fagus) trees. This genus of 'tinder fungus' has also been found in the leather pouch of the remarkably well preserved, 5,500-year-old ice mummy Otzi, which was discovered in 1991 in the Otztal Alps on the border between Austria and Italy.[xix] Otzi is the oldest know European natural human mummy and his belongings are now preserved on display in the South Tyrol Museum of Archaeology in Bolzano, South Tyrol, Italy. A variety of tools and equipment were found by his body, including a copper axe, a flint-bladed knife and a quiver of fourteen flint-tipped arrows. A Stone Age fire-lighting kit composed of flint, pyrite and tinder fungus were also found, confirming that the stone-percussion method was still in use during the Neolithic and Bronze ages. Several known examples of strike-a-light kits have recently been discovered within prehistoric mortuary contexts in mainland Britain.[xx] New radiocarbon data suggests that the practice of depositing a marcasite and flint fire-lighting set in the grave of adult males as symbolic inclusions reached a peak between 2000 and 2200 BC, dating back as far as 2500 BC.

Pyrite and Marcasite

Pyrite and marcasite are minerals usually found in sedimentary calcareous (limestone, or calcium carbonate) rock layers where flint is also normally found. These minerals may also be found in secondary deposits such as beach gravels, in proximity to eroding limestone cliffs, the White Cliffs of Dover (UK) being a good example. The beaches in Folkstone (Kent, UK) are renowned to yield nodules of marcasite as well as a vast array of small marine fossils. In this respect, flint and pyrite naturally occur in the same locations. In some locations, however, several layers of flint nodules are interspersed by layers containing marcasite inclusions. Our ancestors would have likely, and conveniently, foraged for flint and marcasite nodules from the same sites.

Pyrite and marcasite produce relatively cool (less than 600°C) but long-lived sparks when struck, or rubbed against another piece of the same stone or a piece of flint. Pyrite (from the Greek, 'on fire') and marcasite are different crystalline forms of iron sulphide (FeS_2), in admixture with other minor elements and inorganic compounds. X-ray diffraction may be used to differentiate the two forms with ease. In contrast to pyrite, marcasite exhibits rhomboidal crystals which give this mineral a typical round nodular appearance. These nodules are less dense than the cubic lattice of pyrite and display a most characteristic radial structure originating from the centre of the concretion. Because of the radial, aggregate-like nature, marcasite is more reactive to oxygen than pyrite and generally weathers quickly once exposed to the air. The nodules are normally found covered with a thick layer of brown iron oxide, which often gives them the appearance of an iron-rich meteorite, for which they are sometimes mistaken. The layer of oxide protects the nodule from oxidation to sulphates and iron oxide. Once a nodule is split open, it will degrade and oxidise within a few months depending on local humidity and temperature conditions (Figure 13).

Figure 13. Intact and split nodules of marcasite.

Despite this, some intact nodules have survived archaeologically for thousands of years in pre-Roman graves where a male warrior body was normally found along with an early stone percussion fire set. Like pyrite, marcasite may be found across many sites in Northern Europe, including the cliff beaches of England and France and inland clay and limestone deposits in Denmark and Germany. Round nodules of 5–10 cm diameter is what early humans would have looked for, as nodules of this size would have fitted comfortably in the palm of the hand when struck with another stone. The geological formation of marcasite and pyrite in limestone deposits is well documented, and relies on the action of anaerobic sulphate reducing bacteria, which produce hydrogen sulphide. The latter reacts with iron-rich minerals to form iron sulphides. Sulphates are widely available in sea water and marine sediments, as well as iron-containing minerals. The ongoing process of pyrite formation is easy to observe on most modern sandy beaches. A dark, smelly layer of iron sulphide-rich sand can be encountered at a depth of about 0.5 metres. The smell is due to volatile hydrogen sulphide which resembles the stench of rotten eggs. Water-logged compacted sand is the ideal environment for this chemistry to take place, as the hydrogen sulphide eventually reacts with iron-rich minerals to form iron sulphides. On a geological timescale, these compounds eventually crystallise into discrete nodules of marcasite or pyrite aggregates which keep growing until certain conditions are met. The relatively 'crumbly' nature of marcasite, compared to the more solid pyrite, is in part responsible for its superior reactivity to air and, ultimately for its pyrophoricity.

The geological formation of pyrite and marcasite is remarkably similar to a much faster chemical process which sometimes occurs on the inner surfaces of oil tanks aboard cargo ships, where rust formed by corrosion reacts with the hydrogen sulphide gas released by the crude oil to form a mixture of pyrophoric iron sulphides. These can cause explosions when they are suddenly exposed to air during the unloading of the tankers. The same sulphides have also been detected in chemical plants and gas lines, where they pose a high risk of explosion. Research is ongoing to find ways to 'deactivate' reactive sulphides and prevent unwanted ignitions and explosions.

As the chemistry is similar to the formation of pyrite in nature, it will be described briefly here. Rust scale on ships is known to be a mixture of goethite and lepidocrocite, two forms of iron oxyhydroxides. Lepidocrocite $\{YFeO(OH)\}$ will gradually change to the more reactive goethite $\{\alpha FeO(OH)\}$ during exposure to a humid environment. Reaction with hydrogen sulphide will convert the goethite into mackinawite (ferrous sulphide, FeS), which then further reacts to form Greigite, Fe_3S_4 and pyrite, FeS_2. These sulphides are pyrophoric because of their high tendency to oxidise exothermically (i.e. releasing heat), particularly when the surface area of the rusty deposit is increased, as after scraping the surface. The formation of iron sulphides from rust can be described by the following simple equation, which is a similar process to that occurring in marine sediment under anaerobic conditions in the presence of hydrogen sulphide:

$$2FeO(OH) + 3H_2S \rightarrow 2FeS + S + 4H_2O$$

Mackinawite, FeS, may also undergo pyrophoric oxidation via prior partial transformation into pyrite, according to the equations below:

$$4FeS + 4S \rightarrow 4FeS_2$$

$$4FeS_2 + 11O_2 \rightarrow 2Fe_2O_3 + 8SO_2$$

Overall, the process produces iron oxide Fe_2O_3 and releases the poisonous gas sulphur dioxide. The reaction is highly exothermic because of the presence of iron and sulphur undergoing oxidation at the same time. Depending on the size of the particle undergoing oxidation, the heat produced is enough to bring it to self-sustained combustion and hence incandescence. When small particles are detached from a marcasite nodule by flint percussion, these spontaneously oxidise while in flight, aided by the mechanical heat from the percussion blow, and thus ignite. The dull red to orange colour of the emitted sparks indicate an average burning temperature of 650–800°C, which is sufficient to give these particles incendiary properties when they fall onto finely divided, dry tinder.

The mode of discovery of how marcasite can produce hot sparks when struck is uncertain. It is only feasible to think that prehistoric humans made the discovery during the extraction of flint nodules from limestone deposits, where they also came across nodules of marcasite. Early humans extracted flint nodules by hand, often using picks made of antler horn and hard wooden implements. Sometimes these rudimentary tools are still found in abandoned flint mines during archaeological excavation of late Palaeolithic sites. At some point in time, early flint miners must have struck a nodule of marcasite with a piece of flint, and possibly observed dull orange sparks flying off, or landing on tinder which started to smoulder. As in the case of wood-friction fire lighting, the exact circumstances of the first discovery are unfortunately lost in the mists of time. However, once the discovery was made, it is certain that the news must have quickly spread across long distances with early forms of trade and communication between tribes and villages.

Flint

Flint is a siliceous rock related to chert and quartz. It is often found as large nodules in sedimentary rock beds like chalk (calcium carbonate) deposits. The rock is formed over millions of years by the transformation of dead molluscs and crustaceans which become silicified. For this reason, flint found in the south of England and Northern France often contain small, perfectly fossilised marine organisms. Flint nodules occur in different colours (black, grey and green) depending on the presence of trace minerals in the silica (silicon dioxide – SiO_2), but often appear covered in a white crust of carbonate. Flint is one of the hardest stones on the Mohs scale. And when violently struck with another hard stone, it usually fractures into smaller pieces with very sharp edges. The razor sharpness of the new edges allows the stones to be used as cutting, shaving or scraping tools. During the Neolithic period, flint scrapers and axe heads were widely spread across Europe. Even today, flint 'knapping', the art of manufacturing reproduction prehistoric scraping tools, arrowheads and axe heads out of large pieces of flint, is gaining momentum as a new hobby. However, in the Neolithic period, the craft was an essential skill that all men and possibly women learnt as adolescents and widely practised throughout their lives. As anybody who has attended a flint knapping course can testify, mastering the technique takes considerable time. Excellent illustrated flint knapping books are available but only practice makes perfect in this ancient craft. After much practice, it is possible to achieve great control with the hammer stone and produce intricately elaborate tools which are small works of art in their own right. Rather than using a 'favourite' flint tool to strike marcasite nodules, it is most likely that flakes and fragments from flint processing works were used as 'hammer stones' on a very short-term basis. New criteria has recently emerged to identify 'expedient fire-lighting tools and technology' in the late Palaeolithic period.[xxi]

Just like the late Palaeolithic marcasite nodules, which show evidence of being used as strike-a-lights, (particularly the nodules found at Chaleux in Belgium and other sites in Europe)[ii], many finger-shaped flint tools with rounded edges have been found (often alongside marcasite nodules) in archaeological contexts, suggesting that these may have been used to strike sparks out of marcasite nodules. All these tools exhibit a very similar wear pattern, consisting of rounding at one end with microscopic impact fractures due to repetitive striking and/or forceful rubbing against another hard surface (Figure 14).

Figure 14. Replica Palaeolithic flint striker, showing rounded edge.

The scientific literature shows excellent examples of this damage with drawings and microscope photographs. A few of these rounded tools have been found in caves, in proximity to fire hearths which show clear signs of burning, i.e. charcoal deposits and heat-damaged stones. Experiments to recreate the wear pattern using replica flint tools and freshly excavated marcasite nodules have shown that the wear pattern obtained on the modern flint tools was identical to that observed in archaeological specimens, pointing to the authenticity of these early finds. But how was the 'stone lighter' used, and what materials were most likely chosen as tinder to catch the feeble sparks?

Flint on Marcasite

We may assume that early humans would have developed a personal technique for lighting and kindling fire. However, as we can observe from modern experiments and re-enacting practice, the basic actions required to generate sparks with a flint and marcasite set are well documented. To produce sparks, the striking end of the flint do not have to be razor sharp. A smooth, virtually rounded flint surface can be more effective at removing smaller iron sulphide particles from the marcasite nodule. As mentioned above, a marcasite nodule's inner structure seems solid to the naked eye, but in reality it is composed of millions of thin needle-like crystals radiating outwards from the core. The process of producing a spark involves crushing one or more of these crystals at the same time producing enough heat to initiate the oxidation reaction and thus, 'ignite' them. The rounded end of the flint increases the striking/rubbing contact area, which in turns crushes and detaches more crystals, increasing the number of sparks formed.

So, assuming that the crushing mechanism is the most efficient method to obtain sparks from marcasite, a right-handed person would take a split marcasite nodule in the left hand facing the exposed surface, which is placed directly on a piece of suitable tinder. Then the flint 'strike-a-light' tool is held in the other hand as if one holds a pencil, i.e. made to rest on the lower portion of the middle finger and the two middle portions of the forefinger, and the thumb (Figure 15).

Figure 15. Replica Palaeolithic strike-a-light set, complete with amadou tinder to catch sparks from the marcasite nodule.

The user then forcefully rubs the glittering split surface of the marcasite nodule with a short scraping stroke of the flint and repeats this action several times. Occasionally, a spark will be produced among the grey dust which will also scrape off the surface of the nodule, as the latter wears away. This dust should be removed from the tinder, as it quickly clogs the pores and reduces the chance of a spark to become lodged and ignite it. With this technique, it is only a matter of seconds when a spark, or maybe several, will set the tinder aglow. Smouldering ignition can easily be detected as a tiny glowing spot of less than 0.5 mm in diameter, emitting a faint wisp of white smoke. As marcasite sparks are very small and cooler compared to those generated by a steel or a ferrocerium striker (see later), they can be best observed in the dark as they are almost invisible in bright daylight.

Once the tinder has 'caught a spark', skilful and gentle fanning of the tiny ember will grow it to a point where enough heat is generated to set a tinder bundle alight. An ember 'extender' may be used to expand the glowing mass at this stage. The tinder bundle is any small heap of dry, finely divided flammable material, like thin twigs, dry leaves, dry grass, pine needles, shredded bark, that can easily transition from smouldering to flaming combustion. As discussed above, successful coaxing of the ember into flame takes practice to achieve. If the tinder bundle is compressed too tightly, it will inflame with difficulty, but if it's packed too loosely it may fail to accumulate enough heat to inflame. Likewise, if it is damp it will inflame with difficulty, perhaps only after producing much smoke. So, given these challenges, what materials did early humans use to catch sparks from marcasite nodules? The archaeological evidence strongly points to the use of 'tinder fungus' during the Palaeolithic up to the Bronze Age.

Tinder Fungus

'Bracket fungi', which belong to the family of the polypores, are long-lived (up to 40 years) woody, hard, shelf-like fungi that grow directly attached to the bark of certain species of trees. *Fomes Fomentarius* is possibly the best-known example as its use as tinder throughout history is well documented.

Figure 16. A mature specimen of Fomes Fomentarius fungus, cut in half to reveal the upper trama layer.

This fungus grows in many regions of the northern hemisphere, preferably on dead or decaying birch and beech trees, and is commonly known as 'hoof fungus' for its resemblance to a horse's hoof or, quite aptly, 'tinder fungus' or even 'false tinder fungus'. 'Real' tinder fungus is another polypore species, *Inonotus Obliquus*, or more commonly 'Chaga' which infects mostly birch trees as a callous black growth. Like

Fomes Fomentarius, Chaga can be processed into a spongy, soft material that catches sparks effectively from marcasite and keeps smouldering for a long time. When sectioned vertically, the fruiting body of the fungus appears divided into three well distinct sections. The uppermost hard crust covers a dense, spongy layer of suede-like brown material composed of an intricate pattern of intertwined micro-fibres, known as the 'trama', followed by a thick layer of sporing tubes which are not useful to make fire. The trama layer, which is also known as 'amadou' (from the Latin French 'lover' because it easily ignites, and from which tinder is made) can be isolated from the crust and the sporing tubes by various techniques (Figure 17).

Figure 17. Processed amadou layer, after extraction from the trama layer, under the outer crust.

Generally the crust can be removed with a well sharpened knife and a good deal of patience, thus exposing the trama layer. This can then be sliced horizontally to obtain 'sheets' of amadou, which can be used directly after drying as tinder to catch sparks. In some cases boiling in an alkaline suspension of wood ash in water, or dilute saltpetre (potassium nitrate) may turn poor specimens into effective tinder. In all cases, the preparation of good amadou is a skill that takes practice to master. Depending on the micro-fibre density of the amadou and hence its ability to catch sparks, it may be useful to scrape and 'fluff up' the surface with a knife, or shave off a small pile of loose fibrous material, which usually catch sparks much more effectively (Figure 18).

Figure 18. 'Fluffed up' amadou, showing the intricate fibre network.

Once a piece of amadou has caught a spark and starts smouldering, it may burn quite hot for hours without transitioning to flaming (the smouldering front burns at approx. 500–600°C depending on ventilation, when measured with a thermocouple). In some cases, modern expeditioners have made use of this remarkable property by using a whole large bracket fungus (without the crust) as an improvised camp stove, or to carry fire from one camp to the next.

What is amadou made of? Like all fungi, the biochemistry of *Fomes Fomentarius* is complex and involves enzymatic digestion of lignin-derived compounds derived from the host tree into simpler aromatic compounds, like vanillin.[xxii] In contrast to plants, though, which are mostly formed by cellulose, the chemical composition of the fungal cell walls of bracket fungi is mostly chitin in addition to small amounts of other polysaccharides, like chitosan and traces of glycoproteins.[xxiii] Chitin is an unbranched polysaccharide composed of N-acetylglucosamine. Because of the chemical similarity to cellulose, dry chitin fibres of high surface area ignite and burn efficiently. This property is the reason behind amadou's propensity to ignite with ease and propagate a wind-resistant smouldering front. When observed through a scanning electron microscope, the micro-fibres of amadou exhibit an enormous surface area (an estimate of 1500 m^2/g) which allows optimal diffusion of atmospheric oxygen to sustain smouldering. The same property gives amadou a high absorbency which is exploited to this day by fishermen, who use the material to dry and clean their fishing 'flies'. In Romania, amadou is still used to manufacture soft hats, handbags and other items of clothing, often crafted for the tourist market.

WOOD FRICTION AND FIRE ROLL

Wood-friction fire-lighting methods have been used across the globe for tens of thousands of years. Although the earliest prehistoric wood-friction fire-lighting equipment has not survived the passage of millennia, a few well-preserved early Egyptian bow drills have come to light during archaeological excavations carried out in tombs located in sandy desert locations. Hieroglyphs dating from the VIII dynasty and a complete drill with a charred point, which was excavated at Illahun, are testament to the use of bow drill wood-friction techniques by the Egyptians as late as 2000 BC. In this respect, Egyptian bow drills are certainly among the oldest examples to be found intact in an archaeological context.

All wood-friction methods rely on the same basic principle: muscular-derived kinetic energy is transformed into thermal energy (heat) by friction, which may be defined as the force resisting the relative motion of two solid surfaces. Friction arises from a combination of surface roughness and inter-surface adhesion interaction. Heat is thus generated by the forceful rubbing, sawing or rotating of a 'drill' or spindle of seasoned, dry wood against a wooden board or 'hearth', which can be made from the same or a different species of wood. Prolonged mechanical friction between two sliding wooden objects also results in the formation of partially carbonised (charred) dust, which is forced to accumulate inside a 'notch', a small depression normally carved into the hearth board (depending on the method used). As more heat is produced by continuing vigorous drilling or sawing, the pile of accumulated charred dust eventually heats up above 350°C by a combination of auto-oxidation reactions (above 200°C) and continued frictional heating, it then reaches its ignition temperature starting to smoulder, usually from the core outwards. A small self-sustaining ember (often called a 'coal' in the US) is thus formed in the dust pile. At this point we have 'fire by friction'. All known wood-friction methods essentially rely on auto-oxidation chemistry taking over from the initial source of mechanical frictional heating. After ignition has occurred, mechanical action may be halted as the ember propagates without further application of frictional heating. Skill and practice are essential to determine when to stop drilling or sawing, as the ember is still small and fragile at this stage. If the dust pile is suddenly broken up and dispersed, heat would quickly be lost to the surroundings and the ember would be extinguished. Typically, the presence of a self-sustaining ember in the frictional dust pile is marked by a continuous wisp of white smoke being emitted after drilling has stopped, signalling that ignition has been achieved. Being able to judge when the dust pile is about to ignite or has ignited is a necessary skill to know when lighting fire by friction. Smoke density and dust colour offer very good clues here. Very fine, dark brown-to-black dust is preferable to light-brown, coarse or gritty material. The formation of thick, dense white smoke during drilling is often indicative of successful pyrolysis (charring) occurring within the pile approaching the onset of ignition. Photographs of successful and 'poor' dust piles are shown later.

The choice of wood, its density, moisture content and state of decay are all key factors in obtaining fire by friction quickly and reliably. In general, soft, fine-grained woods of low to medium density are preferable to very hard, dense woods. Resinous woods may be used, but the presence of pitchy resin delays or sometimes prevents the formation of a self-sustaining ember, as thermal decomposition of the resin requires heat and sticky tarry pyrolysis products limit oxygen diffusion within the dust pile. Key to success is also the average particle size of the frictional dust and its degree of carbonisation, which in turn govern the rate of air diffusion through the particles and hence heat conductivity. Unseasoned, 'green' wood is useless while surface-wet wood may be used only after drying by prolonged drilling action. However, if the wood has been submerged in water for prolonged periods (days), the likelihood of success is limited unless thorough prior drying is carried out.

The second and equally critical action of wood-friction fire lighting involves transferring the first small ember from the hearth board into a suitable tinder bundle made of dry vegetable matter. As explained above, this is done to grow the ember to the minimum size and temperature sufficient for transition from smouldering to flaming. The presence of loosely compacted, fine and fibrous material like untreated cotton, cattail 'seed fluff' and shredded grass greatly facilitates the formation of a first flame. Heat transfer coefficients within the tinder nest increase with increasing surface area, thus accelerating ember growth. Primitive fire-lighting enthusiasts refer to this process as 'ember extension'. Several natural 'extenders' may be used which are obtained by grinding pieces of decayed wood, bracket fungi or by pounding the bark of certain trees. The high surface area and low auto-oxidation ignition point of extenders are key to reliably obtaining fire. Once the tinder nest is ablaze and steadily flaming, it can be used to ignite a camp fire of small twigs and gradually larger diameter logs.

Drilling Methods

Although five well-documented variants of the wood-friction method exist, the most commonly cited and best known are:

1. The hand drill, or 'two-stick' drill in which a thin wooden dowel or spindle, as it is often called, is rotated by hand into a notch carved into an earth board.
2. The bow drill, where a (usually) thicker wooden spindle is rotated with the alternating motion of a strung wooden bow made with cordage woven

from herbaceous fibres or thin strips of animal leather or modern cordage.

Each of these techniques is described in some detail below.

The Hand Drill

The hand-drill method has been used all over the world, from Native American and Inuit tribes, to China, Australia, and the African continent. Despite the widespread availability of matches in all of these areas, the method is still used to make fire today by Aboriginal tribes in Australia, and certain tribes in interior African countries, although only for ritual purposes or to demonstrate traditional techniques to tourists. Even today, this technique is widely practised by bushcraft enthusiasts, using natural materials that can be foraged in the wild. It is often described as one of the most difficult but rewarding of all fire-lighting techniques to learn, due to its simplicity and the 'primitive' feel it conveys. Fire lighting doesn't get any more primitive than rubbing two sticks together by hand.

The drill is normally crafted from a straight, seasoned stalk about 7–9 mm in diameter, and from 38–50 cm in length. The base, or hearth, can be anything from a 2.5 cm-thick dead branch to a larger flat piece of wood carefully carved into a small plank. A small depression is carved in the hearth with a knife, a sharp flint, and then a V-shaped notch is carved to 'open' this depression to the side of the board (Figure 19).

Figure 19. 'Burnt in' hand-drill set – elder spindle on wild clematis hearth.

This slot collects the charred, frictional dust from the grinding action of the rotating drill and ensures optimal air diffusion into the dust pile, aiding oxidation and eventually ignition. The drill is normally held upright with its end firmly pointing in the depression, and thus rotated by the hands which are kept flat, while exerting downward pressure to increase friction. Because of the pressure required, the hands gradually move downward towards the bottom end of the drill, at which point quick and coordinated action is required to bring them back to the top to start the downward drilling process again. The action of returning both hands to the top of the drill inevitably forces the operator to stop drilling for a fraction of a second. This in turn leads to heat loss in the charred dust pile, which prolongs the onset of ignition. To obviate this, some modern bushcraft instructors have devised an alternative hand drilling technique often referred to as 'floating', which keeps both hands at the top of the spindle during drilling. To achieve this, the drill is rotated between the flattened palms in the usual way but the hands are made to follow an alternate semi-circular upward motion which prevents them from moving down along the length of the drill (Figure 20).

Figure 20. The floating hand drill method is hard to master but allows the use of short spindles.

This also means that considerably shorter drills may be used. UK instructor Dale Collett, founder of the British Bushcraft School, currently retains an application for the Guinness World Record for producing fire by the floating hand-drill method using a 5 cm-long spindle. Collett manages to obtain embers in less than thirty seconds using the floating technique, and is thus recognised as a world expert in fire lighting using this method.

Regardless of the drilling method used, spindle rotation and applied downward pressure must be continued, without pause, until enough charred dust collects in the notch. When copious amounts of smoke are produced by the dust pile, drilling may be halted to check that ignition has been achieved, as evidenced by continued smoking of the dust pile no longer sustained by drilling action.

Figure 21. Close-up photograph of hand drill ember.

An ember may occasionally be produced without a notch in the hearth (this is referred to as notch-less ember) but with considerably more difficulty. In some case the tip of the drill can also become ignited and start glowing, however this requires considerable force and greater stamina to achieve. Once a self-sustaining ember is obtained, this is carefully transferred to a tinder nest and blown into flame. Certain African tribes who still make fire by the hand-drill method have been observed to inflame the tinder nest by running towards the wind or holding the smouldering mass at arm's length and quickly rotating the arm to allow an air stream through the bundle.

However, the ultimate success of the technique depends on the availability of the correct wood type, technique mastery and weather conditions in general. In very hot and dry climes, the hand drill is more reliable and dependable than it would be in damp and cold climatic zones. Also, although a successful hand-drill set may be fashioned in the wild using foraged materials, modern bushcraft enthusiasts generally harvest green wood that is carefully de-barked and seasoned at home, before being used for the purpose. With regard to the origin of the technique, the archaeological record is very scant of information related to the prevalence of the hand drill during prehistory, however, it is feasible to assume that early humans would have used this method before devising more complex methods such as the bow drill or the fire pump.

Making a Hand-Drill Set Like Our Ancestors

As mentioned above, wet and damp climatic conditions may render this technique difficult to master in some regions of the globe. Given that the weather in the northern hemisphere can be wet and damp, even during the summer period, it is important to select the right wood to make the hearth board and the spindle. In the UK, successful combinations include burdock and elder spindles on willow and wild clematis boards, although other wood types (sycamore, buddleia and horse chestnut) may work well when combined with a good drilling technique and lots of perseverance.

A suitable hearth board is first made by choosing a dry, dead branch of willow or wild clematis which is about 25 cm long. The branch is then carefully split with a knife or hatchet and carved to make a flat, 1.5 cm-thick plank, which functions as the board. The spindle is normally obtained from a straight shoot of elder or burdock stem around 30–50 cm long and 8–10 mm diameter. Larger spindles will rotate in the hands at lower angular velocities, thus reducing friction and requiring more effort to apply constant downward pressure. Any knots and side branches must be removed, as well as the bark with a sharp carving knife, as these will likely damage the skin during drilling and cause blistering. Most elder shoots grown in the previous year prior to harvesting, have a central inner core made of pith, a soft, sponge-like material. During drilling, the pithy core contains air and thus acts as an inner source of oxygen that promotes self-oxidation and hence production of charred frictional dust.

In order to make fire, a small depression is carved in the base with a sharp stone or knife, large enough to accommodate the end of the spindle. Then, the drill is positioned in the depression and swiftly rotated between the palms of the hands while applying downward pressure to generate frictional heat and thus produce a deeper, charred 'burnt-in' depression which perfectly fits the end of the spindle. This process is commonly referred to as the 'burning in' of the set. At this stage, a V-shaped notch must be carved with a knife. Different theories have been advanced on which shape of notch works best to produce 'fast embers'. An angle of 90–100 degrees starting from the centre of the burnt-in depression represents an optimum compromise allowing for maximum heat accumulation and effective collection of frictional dust. However, the wood type and its state of decay have some influence on the rate of air ingress inside the dust pile and therefore the likely optimal size of the notch. Anything between 60 and 100 degrees might work if the drilling technique is correct.

Once the notch has been carved, the hearth board is positioned on top of an ember 'pan' (a dry leaf or piece of birch bark) and secured under the operator's knee or foot and the drill positioned in the burnt-in hole to start the hand drilling process. Short drills may be used effectively only by the application of the floating technique, which requires skill and time to master. Longer drills are definitely of more use to novice hand-drill enthusiasts. This is because, as the drilling proceeds, the hands of the operator will naturally work their way down the length of the spindle and will require re-positioning to the top at every cycle. If technique and stance are correct, the first appearance of white smoke should occur within 20–30 seconds. At this point it is important to keep the momentum in order to fill the notch with the much valued charred dust, eventually increasing rotational speed and downward pressure until a thick and steady flow of white smoke is released by the dust pile. Only experience will tell when the operator should stop drilling. In general, once the hand-drill set produces black, fine dust and thick smoke, another 20 seconds of vigorous drilling action suffices to secure a self-sustaining ember, although many variables are at play at this crucial time. Operator's technique and residual stamina, wood moisture content and air humidity all play a role in determining the onset of ignition in the dust pile.

Once a self-sustaining ember is established, it's all down to operator's skill to obtain fire. The ember is carefully transferred into a tinder bundle, made of shredded inner bark or dry grass, and blown into flame. Excellent reviews of the hand-drill technique have been written recently by UK-based bushcraft experts. The reader is directed to these sources for advice and troubleshooting. [xxiv] [xxv] In particular,

Reference 21 demonstrates how a usable hand-drill set may be obtained in the wild and without tools, an essential skill for serious survivalists and bushcraft enthusiasts.

Making a Bow Drill Like Our Ancestors

Making fire with a bow drill set is probably the most common and easiest of all the wood-friction methods to master. In a true survival situation this technique can prove fairly dependable, even without carving tools at hand, as it only relies on a curved, stringed length of wood, the 'bow', to keep the spindle rotating in the hearth board. A small section of hardwood or even a stone with a natural depression in it may be used as a 'hand piece' to secure the top of the spindle and apply constant or variable downward pressure onto it. Because of its effectiveness, the bow-drill method is normally the technique of choice for Boy Scouts and survival enthusiasts worldwide. Unlike the hand-drill technique, the likelihood of suffering hand blisters is minimal. From a historical perspective, and similarly to other wood-friction techniques, the origins of the bow drill are lost in the mists of time. Due to the perishable nature of wood, prehistoric evidence of its use is very scarce, although examples of spindles that may have been used to make fire using a bow technique in the late Palaeolithic period have been documented in literature.[xxvi] In ancient Egypt the technique was certainly used to make fire, as spindle holders made of bone along with wooden hearths with charred friction holes have been discovered after being preserved in sandy tombs for millennia.

As for other wood-friction techniques, the choice of wood and its degree of decay are important factors to consider when building a bow drill set. In general, low to medium density woods work better that hardwoods. Generating an ember with a bow drill made of oak or beech is just about possible but very hard because of the much higher ignition temperature of the frictional dust obtained from dense woods. In the northern hemisphere, and depending on preservation state, several species of trees are known to yield wood of good to excellent quality for bow drill fire lighting. A brief selection includes willow, poplar, horse chestnut, lime, sycamore, cedar and some spruces.

In most cases, spindle and hearth board may be carved out of the same piece of wood. However, some bushcraft instructors prefer to use different wood types in order to prolong the lifespan of the set, or because certain mixed wood combinations may produce an ember much quicker than others. A suitable dead and dry branch of the right size for both drill and hearth, may be found still attached to the parent tree. In other cases, if time allows, it may be convenient to harvest green wood and season it indoors for a few months before carving it down to size.

A bow drill set comprises four essential components (Figure 22).

Figure 22. A small but effective bow-drill set built using materials foraged in the Oxfordshire countryside.

The spindle (or drill), the hearth board, a strung bow and the hand piece (or hand 'hold'). What follows is a simplified set of instructions to obtain a workable bow drill set in the wild using a good carving knife or small hatchet. Although bow-drill sets made with much smaller sized components than those described here may work well, a novice is generally advised to start practicing with a standard size set, which makes obtaining the first ember a much easier endeavour. As in the case of the hand drill, it should be remembered that the spindle diameter determines the number of rotations completed at each stroke of the bow (depending on the length of the latter). This in turn determines the amount of wear on the string driving the spindle and ultimately, the amount of friction exerted on the hearth board. Thicker spindles will rotate slower per bow stroke, and will require more downward force to generate a charred dust pile.

Spindle: a good, usable spindle (drill) may be obtained from either splitting a few centimetres of thick branch with a hatchet and carving a section of the split wood into a dowel shape, or carefully whittling a smaller diameter branch approaching the required diameter of approximately 2 cm. The length of the dowel should approach 30 cm. A usable dowel may also be obtained from carving a section of a larger branch, maybe the one used to obtain the hearth board. Both extremities of the drill should be carved into sharp ends. Some instructors also recommend carving a deep hollow depression at the burning end of the spindle which rubs against the hearth, as this expedient minimises 'inner drilling friction'. When rotating against the hearth board, the inner core regions of the spindle aren't as effective at generating frictional dust as its outermost regions, due to lower angular velocity and thus reduced frictional heating. Although not strictly necessary to achieve success, this technique reduces the downward pressure required to achieve smouldering ignition. In a survival situation however, 'coring' a spindle using improvised tools (like a sharp piece of flint) may not be an easy task!

Hearth board: a 10 cm-thick dead branch may be snapped off the parent tree and, after ascertaining that the wood inside is dry and well preserved, i.e. not decayed or 'punky', the branch may be split longitudinally with the hatchet (a knife may also be used to the same effect if pounded with another branch). One half of the split branch is then split again into a 1.5 cm-thick straight plank, which is then carved with the knife to make it as flat as possible. A width of about 5–10 cm will allow for multiple burn holes in the same board.

Bow: this is usually obtained from a 3 cm-thick, 1 m long section of a live flexible branch. A Slight curvature is desirable to keep the string in constant tension, thus minimising string wear during operation. Some degree of flexibility in the bow is important to reduce slipping of the string against the spindle and thus early breaking. Natural, woven cordage is particularly prone to breaking through frictional wear, although synthetic fibres can be used as a more resistant string, when available.

String: Cordage made from natural and synthetic fibres like leather, cotton, rayon and nylon all make effective string for the bow-drill technique. Obviously, only natural cordage derived from woven tree inner bark or nettle stem fibres can be made, with much skill, in a real survival situation. Clothing can also be used to make cordage if a knife and excess items of clothing are available. The bow may be strung by piercing holes at each extremity to feed the string in and secure it with knots, or by splitting each end of the bow along its length to create cradles in which to secure the string. The bow may also be carved at each end to produce anchoring lugs for the string to be secured against. This is probably the easiest solution during a real survival situation.

Hand piece: this can be obtained from a short section (10 cm) of any hard, dense wood branch. Oak, elm and beech make very effective choices, as the dense wood minimises frictional wear and prevents the drill from embedding deep into the hand piece, thus reducing unwanted friction. The branch may be split to obtain a comfortable hand grip and a small depression is carved in the flat end to accommodate the top end of the spindle. Sometimes, natural stones of unusual shape may also be found, which function well as hand pieces.

Figure 23. A natural concave flint nodule, which can be used as a hand piece.

Figure 23 shows a piece of flint with a natural cavity and a good ergonomic hand grip that the author found in a local forest. Large animal bones could also be made into excellent hand pieces for the fire bow drill, but time and skill are required.

Technique

The position and stance required to operate a bow drill are very different from those necessary for the hand-drill method. Assuming the operator is right handed, the bow is held with the right hand, with the left hand wrist firmly locked onto the shin and the board held firmly on the ground with the left foot, while kneeling over it applying body weight.

Figure 24. Even a very small bow-drill set can be effective with the correct technique and stamina.

The spindle is then 'twisted' within a loop of the bow string, ensuring there is enough tension to drive it without it slipping at each bow stroke. If the bow is slightly flexible, string tension will result in some bending of the bow. If the bow is rigid, the correct string tension depends on the type of string. Strings made with natural cordage (nettles, inner bark or other natural fibres) require less tension or else will easily break. Once a depression is carved near the edge of the hearth board, the bottom end of the spindle is rotated in the depression by vigorously driving the bow back and forth horizontally until frictional dust and smoke are produced (this is known as the 'burn in' process). Once the board has been 'burnt in', a V-shaped notch (open to the edge of the hearth board) is carved in exactly the same way as for the hand-drill technique. Now that the burnt-in depression and the drill end are 'matched' and contact area is maximised, the set can be used to make fire.

A dry leaf or piece of bark is normally positioned underneath the notched depression to serve as an 'ember pan' to catch the frictional dust that collects in the notch. With the wrist of the left arm firmly lodged against the shin of the left leg which presses down onto the board, the right arm may begin to stroke back on forth, with the left hand applying relatively little downward pressure onto the hand piece to begin with. As the drill gains momentum and begins to produce smoke and some frictional dust, downward pressure may be increased along with stroking frequency. Some authors distinguish between a 'powder stage' and a 'heat stage' during which striking frequency is maximised to generate the maximum amount of frictional heat, but little more dust. At the powder stage, it is critical to know how to 'read' the dust produced by the set, in terms of colour, particle size and consistency, as these give a good indication of the onset of ignition and the presence of a live ember. The drill may also start smoking in the hand piece, in which case it needs lubrication with anything that may serve to reduce undesired friction there, like mashed green leaves, animal fat or beeswax. Any frictional energy expended on surfaces other than the hearth board results in wasted energy and promotes early arm fatigue. Ear wax has been known to be effective when nothing else is available (the only case of useful exploitation of excessive ear wax that the author is aware of!).

As for the hand drill, different types of powders may be produced with a bow-drill set under varying operating conditions (Table 1). The colour of the frictional dust is related to the degree of carbonisation and pyrolysis sustained by the powder and hence the temperatures reached at the interface between drill and hearth. Brown dust is an indication that insufficient heat is being produced by the set, and therefore that pyrolytic efficiency at the interface isn't optimised. The formation of black and very fine powder, on the other hand, indicates that sufficiently high temperatures (in excess of 300°C) have been reached at the interface. The high surface area and degree of carbonisation maximises the chance of the powder igniting, as heat is retained more effectively and the kinetics of reaction with atmospheric oxygen accelerated. For the same reason, poorly charred and coarse frictional dust piles will not ignite with ease. Although the visual guide below is a good starting point to learn how to use a bow-drill set, there is no substitute for experience. At the time of writing, excellent hand and bow drill fire-lighting courses are being taught by expert bushcraft instructors throughout the year both in the UK and US. Making fire with a bow drill may be learnt much faster after attending a specific course which covers everything from wood selection, notch carving, operating the set and troubleshooting.

Particle size	Colour	Remarks
Gritty	Light to dark brown	Insufficient stroking speed and/or downward pressure
Fine	Very dark brown	Almost ideal conditions but not enough downward pressure and /or stroking speed
Very fine	Black	Perfect conditions – success!

Table 1.
Frictional dust consistency and colour – bow drill.

Figure 25. Perfect bow-drill dust– black and very fine.
The pile has just ignited and an ember is barely visible.

Other variations of the bow-drill method have been documented as being used in various regions of the globe, which also rely on cordage to keep a spindle rotating against a hearth base: the Inuit 'fire strap', which makes use of a wooden mouth-held piece to apply pressure on the spindle, and the 'fire pump' or compound fire drill, which relies on the alternate motion of a fly wheel spun by the repeated vertical movement of a horizontal bar that minimizes the amount of downward pressure required to obtain ignition. The author has successfully experimented with most wood-friction methods and believes the latter to be the quickest to achieve an ember when a good wood combination is selected. Using a large pump drill with poplar spindle and a wild ivy board, the author easily manages to obtain ignition within thirty seconds with minimal exertion.

Figure 26. Large pump drills like this one (poplar spindle on Ivy board) allow the user to make fire with minimal exertion.

Other Wood-Friction Techniques

The 'fire saw', 'fire plough' and 'fire thong' are somewhat different wood-friction techniques which are still used today by some tribes living in dry tropical regions (Congo, Polynesia, Indonesia and the Philippines). Although these techniques rely on the same working principle and methods involving a rotating spindle, i.e. frictional heating and auto-ignition of charred wood dust, they are designed to convert kinetic energy into frictional heat by translational (non-rotational) motion of a wooden hand piece or length of flexible rattan against a hearth board (fire plough and thong) or a split piece of bamboo held in various configurations (fire-saw methods). Although the origins of these methods are known to date back centuries, having been handed down from father to son for generations, the precise geographical areas of first use are unknown. As in the case of rotating spindle techniques, these methods may possibly have originated after observation of smoke being produced during wood-working processes.

The fire-plough technique, which works by 'ploughing' a pointed, hand-held stick into a longitudinal groove fashioned in a wooden hearth, has recently been the focus of much revived interested in the West. This was the result of the Oscar nominated film *Cast Away* starring Tom Hanks in the role of an air-crash survivor who is cast away for years on a remote island where he learns to make fire by friction using the fire-plough method. Similarly, the fire-saw method works by forcefully rubbing a relatively thin piece of wood (usually bamboo) back and forth over a larger split branch (bamboo or other dense wood). Frictional dust progressively collects in the crack, where it also ignites. The fire thong relies on a thin band of rattan terminating with two improvised 'handles', which is repeatedly and quickly slid back and forth under a flat piece of relatively soft wood (balsa) to ignite frictional dust collecting in a small hole drilled in the board. Excellent photographic descriptions of translational motion methods applied in different configurations are freely available on the internet. The UK survival expert Ray Mears has also described them in good detail in one of his recent survival guides.[xxvii]

The Fire Roll

This fascinating and curious friction fire-lighting method isn't widely known. The technique is currently credited as the fastest friction fire-starting method, and has recently been brought to the attention of the media by German survivalist and bushcraft expert Rudiger Nehberg, who demonstrated it during a German television survival programme. From then on, several enthusiasts uploaded internet videos of their own attempts to make fire, significantly adding to the notoriety of the method. Since Mr Nehberg's first TV demonstration, the technique is sometimes referred to as the 'Rudiger Roll' method, although judging from reputable online blogs, it may have originated in the

Far East in the early nineteenth century, although nothing has been formally published about this technique to this day.

The technique involves pressing and repeatedly rolling a small mass of vegetable fibres between two flat surfaces, typically two small planks of wood or split logs. While one plank rests on the floor serving as rolling base for the tinder, the second plank is slid back and forth on top of it by hand, pressing down with 'reasonable force' at the same time. During this action, the mass of fibres is quickly compressed into a cylindrical shape which is repeatedly rolled under pressure between the two planks (Figures 26).

Figure 27. Demonstration of the fire roll method using a cotton ball with a sprinkling of wood ash to increase friction. The mass is pressed and rolled between two oak planks until ignition is achieved.

Although not strictly necessary, the addition of a small amount of wood ash before compaction of the fibres accelerates the onset of ignition, although other finely divided solids, i.e. sand, and even ground coffee beans and flour have been shown to have a similar, although slower, effect. The addition of solid powder promotes and intensifies mechanical friction within the fibrous mass, rather than acting as a chemical oxidant or ignition catalyst. Moreover, wood ash is mostly composed of stable oxides and carbonates which would not normally function as a chemical oxidiser.

The functioning mechanism of the fire-roll method is simple but remarkably effective, relying on compaction of the natural fibres into a dense fibrous mass of relatively low thermal conductivity. During the rolling action, the relative motion of the individual fibres and entrained solid powder (if any was added at the outset) leads to intense, inner frictional heating of the entire fibrous mass. Given the high thermal insulating properties of a compact fibrous network and two wooden surfaces pressing against it, the tinder mass can't dissipate heat efficiently, leading to its core temperature quickly rising above the ignition point of the fibres (approximately 350°C for cotton, dry grass and shredded inner bark). This heat accumulation is similar to the functioning mechanism of an old hay box, or 'fireless cooker' of the early twentieth century. The device utilised heat from partially cooked food to complete the cooking process. Compacted hay and straw were normally used as insulators for the pot. The process typically took three times longer to cook food but saved a lot of energy and fuel.

When attempting the fire-roll method, experience is required to determine when the rolling action may be halted to check for a 'hot spot' forming in the tinder mass. If the method is performed correctly, the fibrous cylinder should feel very hot to the touch (external surface temperatures above 250°C) after applying a few (normally up to 20) roll cycles. At this stage, the mass is gently teased apart, so that more atmospheric oxygen can diffuse into the central hot zone, causing smouldering ignition of the fibres, as normally evidenced by white smoke being produced. Note that if the mass isn't aerated fast enough, the core will quickly cool below the ignition temperature, forcing the operator to continue rolling the tinder mass. When very dry fibrous materials are used, the ember that forms in the compacted mass is generally self-sustaining and does not require fanning to propagate effectively. Transition to flaming may be accomplished in exactly the same way as for other friction techniques, i.e. by extending it into a suitable, larger tinder bundle.

While reliable historical references about this method are practically non-existent in literature, from a modern survival perspective, this technique may be used to reliably make fire only when two sufficiently flat surfaces can be improvised in the field. Rock slates, which have higher thermal conductivity than wood, may be used if higher pressure and speed are applied onto the rolling tinder mass. Improvised flat planks may be crafted from soft wood (poplar, willow and sycamore) if a knife or a sharp stone are available, which may not always be the case in real survival scenarios. In these cases, improvised rotational drill-based techniques may be more dependable to keep warm.

PERCUSSION OF STEEL ON STONE

Until the advent of the Iron Age, percussion fire lighting relied on pyrite or marcasite used in conjunction with flint to produce feeble but effective sparks to ignite tinder. The Iron Age is widely defined as the prehistoric period in the Afro-Eurasian world (the 'Old world') during which the dominant materials to make tools and weapons were indeed iron and steel. Before then, bronze, which was obtained from copper ore smelting, was used. Although meteoric iron (naturally occurring iron in the form of small meteorites) may have been used from time to time for the purpose of fire lighting from 5000 BC, it wasn't until humans developed the ability to smelt iron ore around 1200 BC, that the production of iron and steel tools became widespread throughout Eurasia. During the Iron Age, production of steel for weapon blades and other cutting tools began with the discovery that prolonged heat treatment of smelted iron in the presence of red-hot coals, would produce a much harder and resilient metal than the starting softer iron. We now know that the process involves the formation of an alloy with carbon (the main constituent of charcoal), with variable contents between 0.5 and 1.5% by weight. Higher percentages of carbon give very hard but brittle steels, similar to what we call 'cast iron' today. A modern cast iron casserole would easily shatter if dropped on a hard kitchen surface, in contrast to a steel saucepan, which may only dent and bounce without breaking. Given its hardness, steel, with a high carbon content will also throw a 'shower' of hot sparks when struck violently with the sharp edge of a flake of flint. Not surprisingly, this property has been exploited to make fire throughout the world from 1200 BC until the invention of the friction matches in the 1820s. That's an incredible 3,000 years of 'flint and steel' fire lighting! A wide variety of shape and sizes of steel strikers or 'strike-a-lights' were developed during this period. Designs varied from the simple and small C-shaped tools used in the Roman period, to the most elaborate and decorative heirlooms of eighteenth-century Britain.

Although historical steel relics tend to oxidise and decompose when buried in the ground for centuries, some early examples of fire strikers that had been buried in male warrior graves as symbolic objects have survived, enabling archaeologists to date and catalogue them. Although excellent publications exist on the subject, in the author's opinion the best bibliographic reference remains the richly illustrated classic book *Fire Steels*,[xxviii] which discusses the origin of each design and contains high-quality photographs of almost every known type used all over the world throughout history. In the United Kingdom alone, some twenty different designs have been in use during the last thousand years. Well-preserved examples dating back to the early medieval and Tudor periods are still frequently discovered on the foreshores of the river Thames by 'mudlarks', and by hobby metal detectorists in open fields. The author has personally found a couple of Tudor strikers with his metal detector in a field near Oxford. Although one of them is too corroded and fragile to survive striking with a flint, the smaller 'bull horn' type he found (Figure 28) still reliably produces sparks, after spending 500 years or so in the soil. Classic 'bull horn' styles were used extensively from the Viking period until the early Tudor years.

Figure 28. Tudor fire steels found with a metal detector in a field near Oxford.

Looked at through the lens of a powerful microscope, 'sparks' ejected from a flint and steel set can be easily observed as small dark grey globules of iron oxide admixed with solidified molten iron. Catching sparks in flight is easily done by striking the steel on top of a sheet of plastic. On contact, the glowing particles melt the plastic forming small circular holes, the rim of which traps the particles landing there upon solidification. This method allows the collection of several sparks for direct microscopic observation. The presence of an iron oxide layer on the surface of most globules confirms that impact with a sharp flint edge delivers enough localised energy to the striker, to 'shave off' and heat to incandescence very small slivers of metal (50 to 100 microns across). The particles are detached from the surface with enough energy to be ejected at considerable speed. Contact with air immediately results in partial combustion, which increases core temperatures to above 1,000°C. Visible branching of the sparks is often observed as a result.

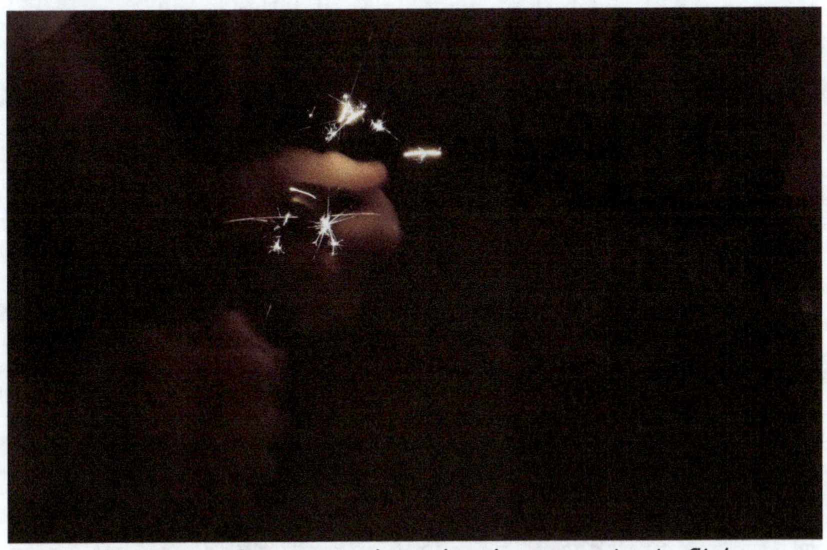

Figure 29. Flint on steel sparks: the extensive in-flight branching is due to burning carbon.

The phenomenon is due to the carbon contained in the particle, which burns to give gaseous carbon dioxide. A 'micro-explosion' is formed inside the molten spherule which splits into several smaller and brighter, hotter particles. The reaction scheme below summarises the process responsible for the formation of branching sparks from impacted steel strikers.

Fe/C surface + hard stone edge + kinetic energy $\rightarrow$ incandescent Fe/C particle + $O_2 \rightarrow Fe_2O_3 + CO_2$ + heat

Although any grade of steel is capable of sparking when struck with a hard stone, harder and tempered steel alloys (containing higher concentrations of carbon) give sparks that burn hotter and branch more effectively while in flight. As this results in increased chances of igniting tinder, a well hardened fire steel was often an indispensable tool, to be found on the fireplace hearth of every dwelling in the land. From the early Middle Ages onwards, flint and steel sets were typically kept in 'dug out' wooden tinder boxes which also contained tinder in the form of tinder fungus or charred rags (char cotton) and, from the sixteenth century onwards, sulphur-dipped wooden splints, or 'spunks', made by dipping thin slices of wood in molten sulphur at one end. Due to the very low ignition temperature of sulphur and its propensity to inflame immediately on ignition, these coated splints would ignite on contact with even the smallest of embers, thus simplifying the task of fire lighting significantly. It could be said that spunks are the forerunners of the friction match, as a chemical element was used for the first time to aid ignition. However, spunks did not contain the vital ingredient that allows modern matches to inflame directly from striking, a 'chemical oxidiser', an inorganic substance capable of supplying surplus oxygen upon thermal decomposition.

Figure 30. Replica oak tinder box from the Tudor period built by the author. Note the square 'damper' which was used to smother the burning tinder, so it could be used again.

In the absence of sulphur, which became widely available only in the late eighteenth century with supplies of volcanic sulphur being shipped from various overseas territories, the small ember obtained by direct sparking of tinder had to be skilfully coaxed into flame by fanning or blowing, as described for wood-friction techniques.

Tinder bags and boxes containing steel, flint and tinder varied in shape and sizes in various parts of the globe during the centuries. From Roman leather pouches, to Tudor 'dug-out' oak boxes and Georgian receptacles incorporating a candle holder, the designs continued to evolve through time to satisfy requirements and the tastes of the different social classes. The highly ornate ivory tinder boxes displayed on the hearth of manorial houses of the gentry, contrasted with the humble dug-out oak types made by cottage dwellers and land workers. A thriving local cottage industry of fire-making tools including tinder boxes, forged fire steels and rush-light holders was thus born in the early Middle Ages. The Bryant and May collection of fire-making appliances, now housed at the Science Museum in London, incorporates hundreds of well-preserved examples of tinder bags and boxes from all periods of recorded history. In the Orient, fire steels were often incorporated within an ornate or plain leather pouch used to carry a small flake of flint and some tinder. Tibetan 'chuckmucks' (Figure 31) are perhaps the most elaborate examples of Eastern craftsmanship as applied to fire-making accessories. Over fifty chuckmuck examples are included in the Bryant and May collection, demonstrating that style and size had also varied in the East over time.

Figure 31. Original 'chukmuck' from the author's collection. This item was bought for little money from a local antiques shop. The steel striker and decorating brass plaques were corroded, and required restoration.

From the sixteenth century onwards, mechanisation of the manual flint and steel technique, resulted in the invention of ingenious wheel-lock tinder boxes and 'tinder pistols' which, just like a flintlock powder pistol, relied on the action of a coiled or wedge spring to activate the device. Charred rag or black powder held in a pan, caught the resulting sparks. Good examples of specialised tinder pistols, some even incorporating an 'alarm clock' that would trigger the devise to light a candle, are described in detail elsewhere.[xxix] Most simple tinder pistols of the eighteenth century incorporated a cylindrical brass or wrought iron candle holder, which would serve to keep a flame alight once fire was obtained. Good examples of British-made early (seventeenth and eighteenth century) tinder pistols have survived to this day. [xxx]

Figure 32. English flintlock tinder pistol dating from the 18th century.

From early medieval times, when the weaving cottage industry grew across England and Europe and the availability of wool and linen cloth became widespread, the preferred source of tinder for fire lighting became charred rag (charred cotton), which replaced natural but less available options like polypore fungi and charred hay. From the Roman period until the invention of the friction match, char cotton was easily made by partially carbonising old cotton or linen rags inside small cooking pots or sheet metal canisters, or simply covering pieces of cloth with hot ashes for a few minutes (which would lead to an inferior but still usable product). The action of very strong heat on the cloth in the absence of air would lead to pyrolysis and carbonisation of the fibres, as opposed to burning, resulting in a two-dimensional sheet of carbon fibres with very high surface area and thus low ignition point. The weaving pattern of the original cloth was retained in the form of a flexible black carbonaceous fabric which could still be folded and torn into pieces for convenience of use and storage. Given the very high surface area and carbonisation degree, even a few dim metallic sparks landing on it would easily set it to smoulder. The tiny ember would easily propagate unaided in a piece of good quality, dry charred cloth. Because of the low hygroscopicity (propensity to absorb moisture), charred cloth would function even in damp and cold conditions, such as those commonly encountered inside non-heated medieval cottages on a rainy winter day. Some early tinder boxes also contained a wooden 'damper' (see Figure 30) which was designed to smother the smouldering piece of charred cotton, which could then be used again. In this way, a single batch would last the household for several weeks. Alas, despite the well-proven reliability of flint and steel to make fire in damp conditions, char cloth would still occasionally refuse to catch a spark due to factors including poor striking technique and degree of fibre carbonisation. The early morning tribulations of Georgian kitchen maids to light the range with 'poor tinder' are well documented. Excessively carbonised cloth ('overcooked' cotton rag that had been held in a fire for longer than necessary) was known to be an inferior source of tinder for flint and steel. This is because, in contrary to common intuition, the removal of all 'volatiles' (pyrolytic gases) from the cloth results in a product with a higher ignition point and lower propensity to catch sparks.

PNEUMATIC COMPRESSION – THE AMAZING FIRE PISTON

The 'fire piston', 'fire syringe' or 'handheld quasi-adiabatic compression igniter', to use its scientifically correct name, was probably first developed in Southeast Asia (Indonesia and the Philippines) in the eighteenth century. The device is arguably the most fascinating and ingenious of all early fire-lighting techniques, due to its ease of operation, quick lighting action and remarkable simplicity. It consists of a handheld hollow tube (which originally would have been made of bamboo, horn or bone) that is sealed at one end and lubricated inside (pig or dog fat were popular with early Oriental versions). The second component is a drift or plunger (the 'piston') which fits snugly into the hollow tube, and slides almost all the way to the bottom. In some early examples, a wrap of greased jute twine was applied to the piston as an improvised sealing ring, as the plunger should not allow air to escape from the tube when it is placed into the bore. In this way, when the piston is quickly pushed in the tube with a single stroke of the hand, the air trapped inside is rapidly compressed and, as a result of this action, is 'quasi-adiabatically' heated (i.e. almost without the exchange of thermal energy to the surroundings) to temperatures exceeding 400°C. In essence, a fire piston may be considered akin to a manually operated single-stroke diesel engine in which mechanical work is transformed into hot compressed gas. A hole carved in the head surface of the piston allows for a small amount of tinder to be introduced and ignited by air compression, without needing sparks.

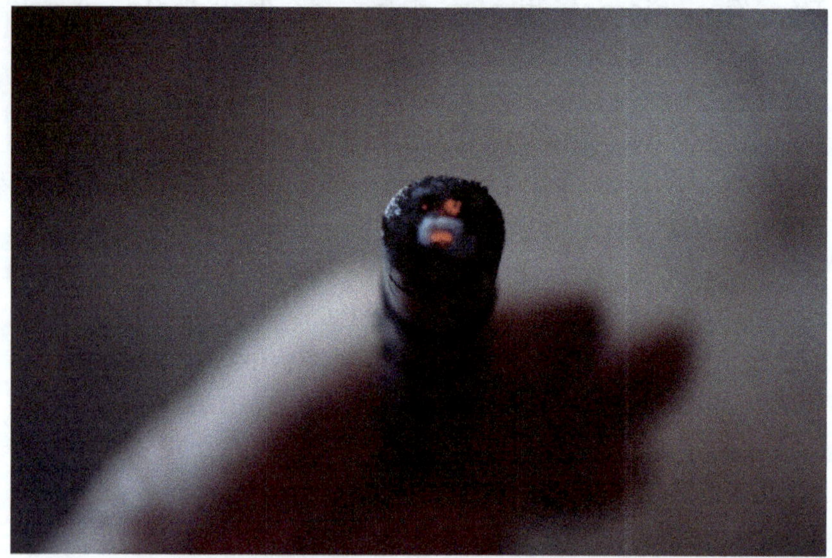

Figure 33. Small ember obtained with charred rag.

All that is required is air and a fast elbow! If successfully operated, when the piston is retracted, the tinder re-emerges sporting a small ember, which can be prized out and transferred into a tinder nest for blowing into flame. Many sources of tinder may be used effectively: dry tinder fungus, char cloth, dry paper or grass and even 'punky' (rotten) wood. As with the origin of the technique, it is believed that accidental smouldering ignition of some wood dust must have been obtained serendipitously during the preparation of a blowpipe bore. Whoever discovered it surely didn't know much about Gay-Lussac's law and the adiabatic compressibility of non-ideal gases.

A metal version of the fire piston was developed and patented in the 1807 in England, following the discovery by explorers who found it widespread across Indonesia. Although several examples of household fire pistons were developed for the British market, the device never really took off as a home fire-lighting tool, possibly because it was tricky to operate in a hurry, and because its functioning still relied on the availability of dry, suitable tinder. Also, the technique still required the operator to grow a small ember and eventually obtain flame indirectly, by blowing onto a smouldering tinder nest.

SOLAR (OPTICAL) METHODS

Sunlight consists of nearly 50% visible light, 45% infrared radiation (IR) and small amounts of UV and other forms of electromagnetic radiation. The invisible IR component has a longer wavelength than visible light and extends from approximately 780–1,000 nanometres. Despite the greater penetrating ability of IR radiation, light of any frequency will heat an absorbing solid or liquid surface, while the rate of heating measured at the surface is proportional to the intensity of the radiant flux. Absorption converts light energy into heat by stimulating molecular motion within a solid surface. Black or dark surfaces heat up very quickly because they absorb all frequencies and reflect none. This is the reason why a black car will heat up considerably faster than a white one on a hot summer day. Fortunately for us, however intense it may get in summer, the average solar radiant flux is not sufficient to set objects (including us) on fire. However, when sunrays are harvested and concentrated purposely or accidentally by optical means, the radiant flux power rises exponentially, easily generating temperatures sufficient to achieve smouldering ignition of most organic materials. With a strong enough flux, flaming ignition is instantaneously achieved.

Most school kids know only too well that it is possible to light a fire using a magnifying glass on a sunny day (Figure 34).

Figure 34. Charring wood with a burning lens.

The author was no exception, and at age eight, almost set fire to the curtains in his parent's lounge on a summer day in Milan, Italy, while experimenting with a small reading glass he casually found lurking in a cupboard. Needless to say, the lens was immediately confiscated never to be seen again. For this technique to be successful, however, the trick is knowing how to correctly focus the sun's rays onto suitable tinder. With a convex lens (a burning lens) this is best done by aligning the plane of the lens perpendicular to the sun's rays while positioning it at its focal distance from the tinder. However, sunrays can also be successfully concentrated (focussed) with a concave reflective surface which may be made of metallic or reflective glass (mirror-like) surfaces. This is known as a burning mirror.

Burning mirrors are also routinely used for cooking and heating water in remote areas of the globe where electric power isn't available. A metallic bracket supports a pot or kettle in proximity of the strong focal point.

Figure 35. Large reflective dish used as a solar heater for boiling water in the Himalayan mountains in Nepal.

A burning lens (sometimes referred to as burning glass) is a convex piece of optically transparent glass or, nowadays, plastic material which can deflect and concentrate the sun's rays onto a very small surface area (usually in the order of square millimetres), thus leading to the formation of a very bright hotspot (the focal point of the lens) where temperatures can reach in excess of 900°C, depending on the focussing power and the strength of the solar flux on the day. When the focal hotspot is impinged upon them, flammable surfaces of wood, paper, cloth and dry grass may smoulder and burst into flame within a few seconds or less. With the development of better glass-making processes in the sixteenth century, large burning lenses were often used in alchemical and early chemical laboratories as a practical, if slightly cumbersome calcination technique, at least until the late eighteenth century. The famous chemist Joseph Priestley used a burning lens to heat substances held inside airtight glass ampoules. These experiments eventually led to the discovery that oxygen ('dephlogisticated air') was required for combustion to take place.

While the shape of a burning lens is fairly standard, a burning mirror may be made of simple polished metal sheet (steel, aluminium, brass) or reflective silver-lined glass of high reflectivity. Burning mirrors may have a fixed or variable focus depending on the design of the reflective surfaces. They may be formed by single, or multiple surfaces joined together at a curved angle.

The heating effect of focussing sunrays with mirrors to make fire was known to the ancient Greeks and Romans, who used concave mirrors to light sacred temple fires in Athens and Rome, where flint and steel were seen as a mundane and impure source of flame. The sun was the almighty life-bearing star which, close to the Gods in the sky, would produce a very pure, sacred flame on earth. Once lit in the temple, the sacred fire would be lovingly tended by male or female priests for months, sometimes years. However, the sacred nature of a solar-derived flame was not always used to connect with the Gods. Archimedes allegedly burnt enemy Roman vessels with his famous 'death ray' during the siege of Syracuse in 213 BC. Recent research carried out at the University of Leicester (UK) to assess whether Archimedes could have really set fire to enemy vessels with polished brass mirrors at a distance of 50 m, found that the patch of reflected sunlight was always greater than the area of the projecting mirror(s), the resulting radiant flux being less that than the incident radiation. As a result of the divergence and the low reflectivity of polished brass, it was concluded that the combined effort of 440 men each wielding a 1 m^2 mirror, would only be sufficient to char and possibly ignite a 0.5 m^2 area of a dry wooden hull at a distance of 50 m. Given the difficulty to achieve this in battle, it was concluded that the classic story may be no more than a myth, although a much smaller army could inflict severe burns by focussing brass mirrors directly onto human targets. Another study conducted in 2005 also independently concluded that brass mirrors would not have been able to concentrate sufficient solar energy to set enemy vessels alight. Interestingly, as a ritual vestige of a glorious past, a polished concave mirror is still used to this day by the chief priestess in Olympia (Greece) to light the first flame of the Olympic torch, which is then carried by hand through the Olympic Games host countries.[xxxi]

Moving to the present time, large burning mirrors and lenses (typically Fresnel lenses, similar to lighthouse lenses) are now used in solar furnaces generally located in remote and arid regions of the world. These are designed to produce high temperatures without electric power. Just like Archimedes' death ray, parabolic arrays of large mirrors focus the sun's rays onto a relatively small area (the size of a car wheel) inside a thermally-insulated cavity, which functions as the furnace, where temperatures can exceed 3,000°C. The fierce heat can be used to melt metal, generate steam for an electric turbine, or simply incinerate waste. With fair weather being the only limitation to using solar furnaces as a source of renewable energy, the largest furnace is located at Odeillo, Mont-Louis in the French Pyrenees, where the sky is clear 300 days a year or more. Solar furnaces are generally known as 'heliostats' or 'power towers' because of the distinctive shape of the sunrays collection tower.

*Figure 36. Medium size solar power station
in a desert location.*

Other well know examples include the Ivanpah facility in the Mojave desert, California, and PS10 (Planta Solar 10), the world's first commercial power tower located near Seville, Spain. At its peak, this tower produces 23 GWh by focussing more than 600 movable mirrors on the tower, where a steam turbine drives an electric generator.

Lighting Fire with a Burning Lens

To start a fire with a burning lens, it is necessary to focus the sun's rays into the smallest and brightest spot on to suitable tinder. Dark, dry organic matter is best for its ability to absorb radiation and rapidly convert it into heat. Charcoal, char cloth, dry dung and dry *Daldinia concentrica* fungi make excellent tinder even for a very small or weak burning lens, such as one obtained in a survival situation from a pair of prescription glasses, a convex piece of ice or a clear plastic bag filled with water. The lens is best held parallel to the surface and brought further in until the smallest focal spot is obtained. Any deviation from this geometry will inevitably lead to a tailed focal spot which is significantly cooler and likely ineffective. The onset of smouldering ignition will be easily detected by the first faint wisp of smoke and the appearance of a glowing, self-sustaining spot on the tinder surface. White paper or other light-coloured tinder may be induced to ignite much faster by first rubbing it with dark soil, as the author serendipitously discovered at the age of eight without knowing why!

Early examples of nineteenth-century tinder and snuff boxes (smoking paraphernalia) sometimes had a small but powerful glass burning lens encased within the lid. The lens could be used to light tobacco or char cloth directly on a sunny day. These days, novelty solar cigarette lighters can be purchased online for a few pennies. They rely on a small concave reflective plastic mirror which can be easily carried in a pocket (Figure 37). Needless to say, they are very good to help one quit smoking on rainy and cloudy days!

Figure 37. Small plastic solar mirror used to light a cigar on a sunny day. Ignition is achieved in less than 5 seconds.

CHEMICAL IGNITION AND FRICTION MATCHES

Since ancient times, sulphur has been used with flint and steel due to its low ignition temperature (250°C). The use of sulphur in fire lighting has been recorded since Roman times and the technique didn't change for two millennia, at least until the friction match was invented in the nineteenth century. To make 'sulphur matches', splints of softwood were partially coated at one end, to form a solid head. Several of these early matches (or 'spunks' as they were known in medieval England) would be stored in the tinder box with a flint and steel set and char cotton. The sulphur-coated end of these early matches was lit by contact with the small and feeble ember obtained by catching sparks on char cotton. A faint blue flame would indicate that the sulphur had ignited and soon enough, the flame would turn yellow, once the wooden splint caught fire too. The stick flame was used to light a candle, a rush light or start a fire in the fireplace. With the possible exception of MnO_2 nodules possibly used by Neanderthals thousands of years before, sulphur matches may be considered the very first example of deliberate chemical kindling, and possibly the forbearers of modern friction matches as we know them today. Another early form of chemical kindling involved the manufacture of a 'slow match' by soaking ropes in dilute solutions of potassium nitrate (saltpetre). A slow match was a slow-burning fuse required to operate musket and cannon until flintlock ignition was introduced in the sixteenth century. As opposed to sulphur, which is a flammable fuel, slow-match manufacture involved the use of an oxidiser (saltpetre) to aid ignition and propagation of the rope, which acted as the fuel. The word 'match' is intimately linked with gunpowder manufacture and pyrotechnics, deriving its name from the Latin 'mixa', the wick of a lamp. The word 'miccia' is still used in modern Italian to indicate a fuse, typically of a firework or other pyrotechnic article ('accendere la miccia'; 'to light the fuse').

The Discovery of White Phosphorus

Despite the use of sulphur as a chemical kindling agent for the best part of recorded history, the very first examples of chemical ignition devices made their appearance after 1669, the year white phosphorus was discovered in Hamburg by the alchemist Hennig Brand while distilling a mixture of urine residue and charcoal in an earthen retort.

Figure 38. Hennig Brand.

As many of his contemporary alchemists, Brand was searching for the philosophers' stone, an element capable of transmuting 'vile matter' into gold. He therefore had the idea of using the 'golden stream' (human urine) for that purpose. He eventually obtained a waxy substance, heavier than water, that would glow green in the dark and would catch fire spontaneously when rubbed between two surfaces. Brand had just accidentally isolated the most reactive allotrope of phosphorus, the 'white' form of the element. Brand called his new substance 'cold fire' and eventually phosphorus (from the Greek 'light bearer'). Although he first kept his method secret for many years, his invention would eventually allow others to produce fire on demand with great ease for the next 400 years. The discovery of phosphorus had a profound effect on the alchemical circles of the time. The waxy substance gave off a peculiar garlic-like smell, inflamed spontaneously with a brilliant white flame and shone in the dark with a beautiful green glow. When the retort distillation method became known across Europe, elaborate demonstrations of phosphorus' 'magical properties' were set up for the dignitaries of early scientific communities, and even for the amusement of kings and queens to great effect.

In 1670, the English chemist Robert Boyle obtained his first samples of the element by repeating Brand's original experiment, but also by heating sodium phosphite with sand (silica). Boyle renamed the substance *icy noctiluca* (cold light) and, for the first time, started experimenting with the element in a scientific manner, examining its properties and reactivity. Boyle also discovered that when sulphur-dipped wooden splints were drawn across a piece of paper bearing traces of phosphorus, the sulphur tip would ignite at once without friction. Although this may be considered the very first recorded example of an oxidiser-free friction match, Boyle's invention remained a chemical curiosity for many years without leading to any commercial product. It would take another 100 years for usable fire-lighting devices based on white phosphorus to appear on the market. In 1781 the phosphoric taper (or phosphoric candle) was introduced in France.[xxxii] It consisted of a 10 cm sealed glass tube containing a small piece of partially waxed paper and a small globule of white phosphorus at the bottom. After immersion of the tube in hot water, the phosphorus globule would melt and impregnate the non-waxed bottom end of the paper. To make fire, the tube would be broken (sometimes with bare teeth!) and the taper quickly withdrawn, whereby it would immediately burst into flame in contact with the air. Phosphoric candles were too expensive and never really took off as a practical fire-lighting tool. However, a similar contraption, the 'phosphorus bottle' or 'pocket luminary' was developed only a few years later in Italy. It consisted of a small airtight glass flask or ampoule containing a thin layer of white phosphorus (partially oxidised after the vial had been opened a few times) deposited on its inner surfaces, and a number of wooden splints tipped with sulphur. To get a light, one of the splints would be rubbed against the coating in the vial and then quickly withdrawn, whereby the sulphur tipped head would inflame almost immediately. A very similar device, the briquette phosphorique, appeared in France a few years later. Cheaper and more slender than the first phosphoric candles, a number were sold across Europe but never really took off as the phosphorus coating would become ineffective after only a few matches were lit. Some original examples of these early matches (and phosphoric bottles) can be seen in the Bryant and May Museum of Fire-Making Appliances in London.

During the period 1780–1830, a good number of very ingenious fire-lighting devices not containing white phosphorus were developed. Some of these successfully transitioned from chemical curiosity to commercially available appliances, despite the risks involved with the pyrophoric, finely divided metal powders they contained. Pyrophoricity is the property of a solid or liquid compound to ignite spontaneously when brought into contact with the air. Examples of pyrophoric substances abound in the modern chemical laboratory, where highly reactive reagents (often organic compounds of alkaline metals) are routinely used for synthetic work. Back in the late eighteenth century the only substances known to display pyrophoric properties were white phosphorus and finely divided metal powders, especially aluminium and iron. Homberg's pyrophorus was the first pyrophoric device accidentally discovered by Wilhelm Homberg. The device relied on fine aluminium powder prepared by calcining a mixture of alum (potassium aluminium sulphate) with flour in a glass vial.[xxxiii] Upon breaking the vial and scattering the contents, the powder spontaneously ignited. Other examples followed, from Hare's pyrophorus (made by roasting Prussian blue) to Roesling's powder, designed to provide an instant (and arguably dangerous) light for pipe smokers.

Catalytic Ignition of Hydrogen Gas

In 1823, the German chemist, Professor Johann Wolfgang Döbereiner invented the 'hydroplatinic lighter' or 'Feuerzeug' which relied on the catalytic ignition of hydrogen gas over platinum 'sponge' (a finely divided and porous form of the metal). Hydrogen was generated in a glass jar similar to a modern chemical Kipp's apparatus, by the reaction of zinc metal with dilute sulphuric acid. By opening a small brass pneumatic valve, a tiny stream of hydrogen gas was made to impinge into a brass cup containing platinum sponge, which catalysed the highly exothermic combustion reaction of hydrogen with atmospheric oxygen. The heat released was sufficient to bring the spongy catalyst to red heat thus igniting the hydrogen stream, producing a small, very hot but almost invisible flame. A schematic drawing of this device is given in Figure 39.

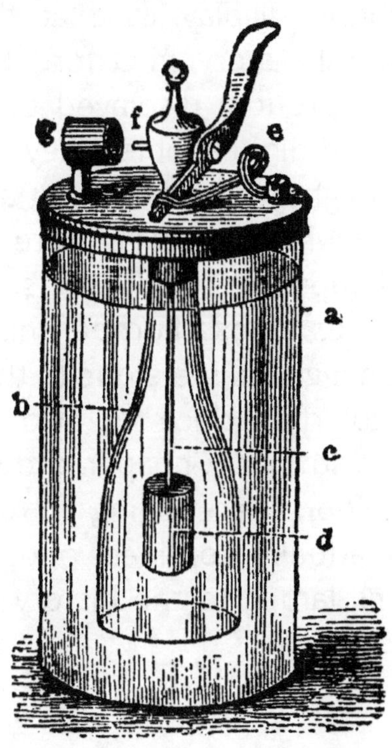

Figure 39. Schematic of the Döbereiner lamp.

During the 1820s and 30s several designs of this early lighter appeared across Europe, despite the inherently hazardous properties of hydrogen gas and dilute acid. The lighter was prone to explode if air was allowed to leak back into the jar. An exploding lamp would produce a shower of glass shards and a splatter of dilute acid droplets, potentially causing serious harm and damage to clothes and furniture. Despite the inherent danger, a significant number of Döbereiner lamps were sold to the Victorian elite classes and later examples featured ornate and richly decorated cases showing classical themes to disguise the inner glass jar. Fabulously ornate Victorian examples can be seen in the fire lighting display case at the Pitt Rivers Museum of Natural History, in Oxford, England. These lamps would be proudly displayed on the fireplace mantelpiece or the dinner table. They would be used to get an instant light for smoking, but also for lighting tapered candles which in turn, were used to light paraffin (oil) lamps. Despite the sale of several thousand examples across Europe, this device never really took off on a global scale due to the hazards and the need to regularly replace the acid solution which became weaker with use. Cheaper and safer methods of chemical ignition were being introduced in the 1830s which gradually confined phosphoric bottles and Döbereiner's lamps to the history of fire-lighting devices.

Electrical Ignition of Hydrogen Gas

Early scientific investigation of electricity was well established by the early 1820s. The Voltaic pile, the forerunner of modern batteries, was invented by Italian physicist Alessandro Volta in 1799. Volta had shown that electricity could also be generated chemically and, like many of his contemporary physicists, had also shown that small electrical sparks ('electric fire' as it was called) could be generated at will by the charging and discharging of early forms of condensers. Volta had noticed that these sparks could ignite the methane gas bubbling out of swamps and later, hydrogen gas. Following these discoveries, hybrid devices relying on electrostatic (spark) ignition of hydrogen gas made a brief appearance during this period. The 'electropneumatic' lamp was invented by Volta himself. This device relied on a tiny electrostatic spark generated by an electrophorus (a simple manual electrostatic generator) to ignite hydrogen gas generated by a zinc metal/dilute acid tank. The electrophorus relied on rotating solid sulphur, wax or mica discs to collect a charge. Similarly to Döbereiner's lamp, the device was hazardous due to potential air suck-back and explosion of the jar.

New Oxidisers

The discovery of potassium chlorate, a powerful chlorine-based oxidiser in solid powder form, is credited to the French chemist Louis Berthollet, who, in 1823, first obtained it by bubbling chlorine gas into concentrated potassium hydroxide solutions. When a stoichiometric mixture of potassium chlorate and a finely divided organic fuel (charcoal, sawdust, fine sugar) is subjected to intense frictional heating, the mixture ignites and readily supports a vigorous flame. The chlorate first melts (~202°C) then decomposes releasing the oxygen required for combustion of the fuel. This property was soon recorded by Berthollet himself who started experimenting with different fuels including sulphur, which is now regarded as chemically incompatible with all chlorate salts. Mixing potassium chlorate with sulphur often resulted in powerful explosions rather than controlled ignition, as Berthollet discovered to his detriment. Several deaths were recorded in the 1820s following experiments where large amounts of chlorate were accidentally or deliberately mixed with sulphur and white phosphorus. It was also soon discovered that a mixture of potassium chlorate and sugar (or charcoal) could be brought to ignition by mere contact with a drop of concentrated vitriolic acid (sulphuric acid), thus not requiring frictional heating to start the reaction. At the same time, other workers, often apprentice apothecary chemists, started to dabble with very dangerous mixtures containing white phosphorus, sulphur and potassium chlorate in various proportions. The ingredients would be mixed as a wet paste and left to dry naturally. Wooden splints would be dipped in the wet mixtures (with or without a glue) and left to dry to produce early forms of friction matches. Not surprisingly, these were reported to be incredibly sensitive to friction and impact, often exploding with a loud report when attempting to strike them or stepping on them. They were also prone to self-ignition, hence posing a serious fire hazard.

The Great Race for 'Safer Match' Supremacy

Despite all the hazard, early experiments eventually led to the development of much safer matches although most still contained white phosphorus in the head composition. Several varieties were introduced across Europe by different inventors claiming title of having first developed them. John Walker in England, Jacob Kammerer in Germany, Charles Sauria in France and János Irinyi in Hungary all independently developed different match-head compositions. 'Congreves' and 'Lucifers' are typical trademarks of this period. Congreves, named in honour of the rocket pioneer Sir William Congreve, were invented by John Walker, an English apothecary well known in the UK as the inventor of the friction match, although several others had contributed before him. Walker apparently sold thousands of tins of his 'friction lights' from his small apothecary shop in Stockton-on-Tees, England.[xxxiv] His friction lights were made with thin strips of waxed cardboard dipped in a chlorate-based composition and were designed to be ignited by pulling between a folded piece of sandpaper, which was also supplied in the tin. In 1831, the French chemistry student Charles Sauria, who is credited as the inventor of friction matches in France, discovered that sulphur could be substituted with antimony sulphide, which also reacts with potassium chlorate but in a much more controlled and non-explosive manner. His new matches ignited reliably and burnt very smoothly. The French government recognised the importance of Sauria's contribution by awarding him a gold medal.

During the 1820s to 1840s, a number of different match-head compositions were developed across Europe. Most of these, including those sold in England, still contained highly reactive and poisonous white phosphorus. Workers in match factories were often afflicted by 'phossy jaw', a dreadful disease which caused necrosis of the jawbone, as the worst symptom of chronic exposure to white phosphorus vapour. Many young children also died as a result of chewing on these early matches, while countless adults resorted to ingesting match heads as a means to commit suicide. Notably, the toxicity of white phosphorus eventually led to the events of the Bryant and May factory revolt in England, which contributed to the demise of white phosphorus being used as a key ingredient in match manufacturing.

During this period, improvements were made to the physical form of the match wooden splint. Swans Vesta's were invented in England in 1832 by dipping fabric and later wooden splints in molten wax to provide a more reliable ignition and help suppress afterglow. The Americans also hold credit. During the 1830s, a number of US patents claimed improvements in the manufacture of 'friction lights'. The most renowned worker in the field was possibly Alonzo Dwight Phillips of Springfield, Massachusetts, who developed the 'Loco-Focos' matches. Instead of the standard composition at the time containing white phosphorus, potassium chlorate, antimony sulphide and gum arabic as the binder, Phillips developed a thick paste containing potassium chlorate, chalk, glue and only trace amounts of white phosphorus as ignition sensitiser. Thin pine splints were first dipped into molten sulphur and then, once the sulphur head had solidified, they would be dipped again in the new friction-sensitive composition. Phillips' matches were also packed in a way to avoid accidental self-ignition for the first time.

In England, during the 1840s, 'Fuzees' and 'Vesuvians' were invented as a type of outdoor, windproof match. Like the American Loco-Focos, they relied on multiple layers of different compositions, achieved with successive dipping cycles. The first coating was a mixture of powdered charcoal and/or fine wood sawdust with saltpetre (potassium nitrate) and hide glue. Once dry, the head was dipped again into standard ignition composition. These matches produced a steady windproof smouldering ember rather than a flame, and soon found use as reliable igniters for railway signals, and for lighting firework displays. Fuzees were still produced by Octavius Hunt in the UK until 2015, when the production line was finally closed down. (Figure 40)

Figure 40. Modern Fuzee matches imported from China.

Finally Safe:
The Discovery of Red Phosphorus

The key event that revolutionised the match industry worldwide was the discovery of the much less reactive red allotrope of phosphorus by Anton Schrötter in 1845. Schrötter had shown that white phosphorus could be converted into a more stable red form by the prolonged action of heat and light in the absence of air. The red form was much less reactive due to its chemical structure, but still reacted with potassium chlorate when a mixture was subjected to intense mechanical friction, suggesting it could be used as a safer alternative in match manufacturing. Arthur Albright in England eventually developed a viable mass production process to manufacture red phosphorus cheaply enough to use it in the match industry. At the same time, the Lundström brothers and Professor Pasch in Sweden developed a 'safety match' in which the red phosphorus was confined to the friction strip painted on the outside of the box. Because the heads contained only potassium chlorate, various fuels and a binder, safety matches could only be ignited by striking on a red phosphorus strip, thus eliminating the risk of accidental ignition typical of sensitive strike-anywhere (SAW) matches. The high toxicity of white phosphorus was also eliminated for good, but despite this progress, safety matches initially struggled to take on commercially as they were still more expensive than SAW matches. It would take a decade or so for them to take off in Europe and the US, where new types of matches were being invented.

Book Matches

Joshua Pusey was a Philadelphian attorney and science enthusiast credited with the invention of safety 'book matches' in 1889. Pusey called them 'flexibles' as they were made of cardboard strips dipped in safety-match head composition and had luminous paint applied to the outside of the case to make them visible in the dark. The striking strip was positioned inside the folding flap of the case. The Diamond Match Company eventually bought the patent, and later relocated the ignition strip on the outside of the case, thus first developing the familiar matchbook cases which we are all familiar with today. Matchbooks took off during the 1890s in the US and sometime later in Europe as a convenient and cheap form of advertisement. Even today, matchbooks are often given away for free by hotels, theatres, nightclubs and bars for advertising their business, and have become collectors' items for the discerning traveller (Figure 41).

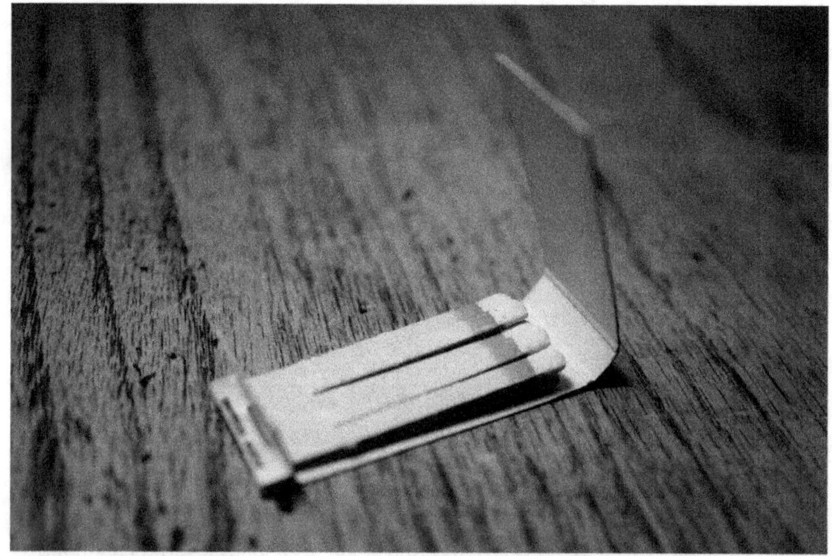

Figure 41. Promotional book match case from a New York night club.

The author's late uncle Tony, a successful insurance broker from Italy, used to keep a big wooden bowl filled with hundreds of matchbooks collected during many business trips around the world. As a young boy, the author was mesmerised by the variety of uncle

Tony's collection which contained brightly coloured matches (gold, pink, green and purple heads dipped on blue and black splints). Unfortunately the entire collection was disposed of after uncle Tony's passing and before the author had a chance to claim possession of it!

During the mid 1850s, research was actively going on to find replacements for white phosphorus in matches. In 1864, Jöns Jacob Berzelius discovered phosphorus sesquisulphide, P_4S_3 (tetraphosphorus trisulphide according to IUPAC nomenclature), a yellow-grey solid compound which he prepared by heating white phosphorus with sulphur in the absence of air. Following the discovery, scientists working for the French government eventually managed to prove that the compound was not poisonous but still reacted with potassium chlorate to cause controlled ignition. Phosphorus sesquisulfide could therefore be used to make a reliable and inherently safe match composition and is still being used in match making to this day. The compound was first patented as an ingredient of safer matches in 1898, when it was found to be non-toxic, very stable and less prone to hygroscopicity in a variety of match-head compositions with varying degrees of friction sensitivity. Today, phosphorus sesquisulphide is manufactured by the reaction of red phosphorus with sulphur. The patent was eventually purchased by Albright and Wilson in England, who first developed a method to produce it on a commercial scale, and went on to sell it to several match manufacturing companies of the time. Despite the significant step forward in safety, white phosphorus would still be used for many years until several countries formally banned its use in match manufacturing between the years 1870–1920. India and China were among the last countries to implement the ban.

Modern-day Matches

Nowadays, modern matches are still sold around the world as the 'safety' or SAW types. The latter are still in widespread use in Europe (see 'Vestas' in the UK or the 'Fiammiferi Familiari' in Italy) and in most African and South American countries (Figure 42).

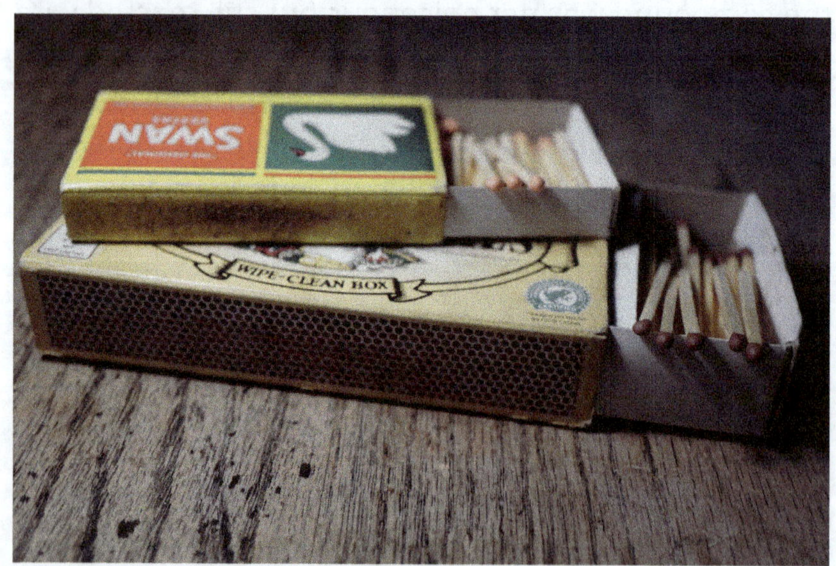

Figure 42. Safety and SAW match boxes sold in the UK – Notice the different strip on the side.

SAW matches are preferred to safety matches in survival kits for their ability to ignite by being struck on any rough surfaces (rock, walls, wood, etc.) and, with some skill, even by pinching between finger nails or the teeth! Despite the widespread use of disposable plastic gas lighters, matches are still mass produced across the world, including developing countries. The relatively simple manufacturing process, coupled with increasingly available wood obtained from sustainably-managed forests and the availability of cheap, mass-produced Chinese potassium chlorate and other chemicals, ensure that matches remain cheaper than disposable butane gas lighters, or other means of obtaining fire. Matches are still manufactured using splints made of waxed paper, paraffin-impregnated wood, or wax-impregnated cardboard (for matchbooks). Notable variants are marketed for very specific uses; longer wooden matches with improved burning characteristics are available for lighting fireplaces and stoves. Reliable barbeque fire-starters are also now available. These are manufactured by dipping the tip of wax-impregnated wood-shaving cubes into safety match-head composition so they can be struck against a red phosphorus strip, akin to a safety match. Some types of small fireworks rely on a friction-type match head for manual ignition (notably port-fires and fire crackers, which are still popular in Italy and the US), and some marine distress light signals (torches and hand-rockets) are also ignited by means of a match-head composition applied in various geometries (ordinary striking strip, friction toggle pull chord or other types, as appropriate).

Match-head Composition Chemistry

The science underpinning match manufacturing lies at the interface between energetic materials chemistry and pyrotechnics. Excellent reviews have been written to describe many technological aspects of match making,[xxxv] including ingredients selection and testing. Akin to other fields of chemical technology, improved match-head and striking-strip formulations are still being developed to this day. The Japanese are active in the field and have patented several 'low odour' and coloured flame compositions in which sulphur is replaced with other fuels (notably various organic sulphides, nitrocellulose and diphenylguanidine). Potassium chlorate may be substituted with barium or strontium chlorate to give a green or red flame when mixed with chlorine-containing fuels in the head composition. These matches seem to have found success as novelty items among young people in Japan and other Asian countries, and are not dissimilar, in chemical terms, to small coloured fireworks. Improvements have also been made to make match heads more resistant to moisture, using a variety of natural and synthetic binders. However, animal-derived glue is still the preferred binder of choice in simpler and cheaper formulations, due to its abundance and low price, despite its moderate resistance to moisture. Head compositions with improved combustion properties (i.e. no sputtering flame or ejected sparks) and increased striking sensitivity may be obtained by incorporation of various modifiers, including silica 'microballons' with a typical diameter of 100 micrometres. Table 2 shows examples of typical head composition formulas which include microspheres as sensitisers. Table 3 shows that safety and matchbook formulations, which have very similar compositions, still contain a significant amount of sulphur despite its potentially explosive reactivity with chlorates. However, when fully dispersed among other ingredients (fuels, binders, burn rate modifiers and pigments) at loadings of 5 weight %, its reactivity is essentially reduced to that of a sensitiser (or ignition promoter). Also worthy of note is that strike-anywhere formulations generally contain less chlorate than safety match-head formulas, making the former just sufficient in oxygen (positively, but not overly oxygen balanced) to support propagation without generating a sputtering or excessively violent flame. However, the presence of some phosphorus sesquisulphide, in conjunction with somewhat higher percentages of zinc oxide (a burn rate modifier), ensure stable propagation of the SAW composition and the generation of enough heat for successful splint ignition. In addition to paraffin wax, wooden splints are nowadays also dipped into a fire-retardant solution to suppress 'afterglow' once the match is blown out. This is necessary for safety purposes, as a discarded glowing matchstick could start a fire if dropped on dry flammable tinder. Typical suppressants are dilute aqueous solutions of ammonium chloride, ammonium phosphate (mono- or di-basic) and borax. Moisture resistant striking-strip compositions based on various polyvinyl chloride 'plastisols' (water-based solutions containing PVC, dioctyl phthalate and chloroparaffins) have been patented during the last few decades. These binders ensure the striking strip remains protected from moisture, thus retaining its integrity throughout the life span of a typical match box.

Ingredient	Weight percentage in dry composition		
	Safety match	Strike-anywhere match	Book match
Potassium chlorate	50	35	55
Animal glue	12.5	17	10
Microspheres	12	19	7.5
Zinc oxide	1.0	6	5
Pigment	2.5	0.5	0.25
Sulphur	5.5	-	5.0
Manganese dioxide	9.0	-	-
Kieselguhr	5.5	-	6.25

Potassium dichromate	1.0	-	0.75
Hydroxyethylcellulose	1.0	-	-
Phosphorus sesquisulphide	-	7.0	-
Calcined gypsum	-	15.5	-
Glass powder	-	-	7.5
Starch	-	-	2.25
Gum Tragacanth	-	-	0.5

Table 2.
Typical formulations for SAW, safety and book matches.[xxxv]

Ingredient	Weight percentage in dry composition
Plastisol mixture	50
Red phosphorus (particle size below 30 microns)	30
Antimony sulphide	17
Fine glass powder	1.5
Ferric oxide	1.5

Table 3.
Improved formulation of a striking strip for safety and book matches, containing red P and plastisol mixture.[xxxv]

Match Ignition Theory

The ignition of a match is regarded to be a complex sequence of chemical reactions initiated by the frictional heat generated during the manual striking process. Following ignition, the first spit of flame is propagated by a series of highly exothermic reactions which eventually cease when all of the fuel/oxidiser mixture present in the match head has been consumed. At that point, in a well-designed and produced match, enough heat should have been delivered to the wooden splint to melt the paraffin wax coating and successfully ignite the wood. If the match head is too small as a result of insufficient or partial splint dipping, or if the head detaches from the splint following ignition, the splint may fail to ignite reliably, thus giving poor quality matches.

Despite almost two centuries of progress in match development, relatively little has been published in literature about the principles underlying the ignition process of a match head. However, pyrotechnic chemistry and thermodynamics come to the aid, as much research has been conducted for military purposes during the last 50 years. Detailed studies and computer modelling results of self-propagating solid- and liquid-state exothermic redox reactions have been published for a range of military pyrotechnic articles ranging from delay compositions to IR decoys based on MTV (Magnesium-Teflon-Viton). Some of this work may be theoretically applicable to the ignition and propagation of a simple match head. While thermodynamics is useful to model heat and mass transfer processes, simple stoichiometric chemical equations are also required to describe the chemical transformation occurring in the match head following ignition.

Starting with thermodynamics, the relation between the speed of reaction (reaction rate) and the heat released is described by a modification of Le Chatelier's Principle. Where the relation between the heat released and the maximum temperature reached by the combustion front in the match head, takes the form of an integral equation which requires knowledge of heat capacity (the amount of heat stored by the unit mass at a given temperature) of each phase present in the reaction zone (solid, liquid or gas). In a burning match head, the total amount of heat released and the burning rate typically depend on the components and stoichiometry of the formulation. The geometry of the match head, which is typically 'pear-shaped' as a result of the dipping process, also takes part in the process. Following ignition, the combustion front liquifies into a viscous syrupy mass which gives rise to 'foaming', a gassing phase where tiny gas bubbles are observed breaking out of the molten, thick liquid mass. These are sometimes described as channels. The degree of foaming for a particular composition, i.e. distribution, shape size, and collapse mode, of these micro-bubbles also take part in determining overall chemical kinetics. These processes can be modelled by extending the above-mentioned equations to account for the foaming phase but several assumptions are required.

Equations are models which only approximately describe the many factors controlling burning rate for a compressed match-head fuel/oxidiser mixture, able to propagate exothermically. However, they do not adequately describe the rather obscure pre-ignition phase, which occurs during the striking action. Mathematical treatment becomes difficult at this point, mainly because of the imperfect geometry of the match head and the uncertainty in identifying the exact location where Ti (ignition temperature) is first reached on the surface of the pear-shaped head during the striking action. This in turns depends on the angle at which the match is held when being struck, and the force used to strike it. A small amount of material is typically lost on the striking strip during striking, complicating matters even more. Table 4 shows a simplified sequence of physical actions and related events occurring when striking a match.

Physical action	Chemical action	Controlled by
Initial gentle striking on friction surface	Abrasion and melting of outer surface protective layer	Nature and thickness of protective layer (if any) and nature of the striking surface (SAW or safety match)
Increased friction, with more downward pressure applied on striking surface	Increase in temperature on surface of match head. Solid state pre-ignition	Speed of striking, pressure applied, formulation stoichiometry and moisture content

Continued friction with sustained downward pressure being applied	Start of incandescent reaction and first appearance of luminous flame. Propagation of hot gases through channels in molten foaming phase. Propagation of solid- and liquid-state reactions	Distribution, shape and size of micro-bubbles and channels formed during the foaming phase. Viscosity of molten liquid phase which depends on binder choice and other variables
Removal of ignited match from friction surface	Incandescent reaction rate increasing until fuel/oxidiser system is consumed	Nature and stoichiometry of formulation. Immediate environment around match head (damp, windy, etc.)

Ignition of splint from incandescent match head	Oxidation of cellulosic and/or hydrocarbon wax components of splint relying on atmospheric oxygen to sustain	Nature of splint (wood, wax-impregnated board, type of wax present). Immediate environment (damp, windy etc.)
Continued burning of splint	As above	As above. This stage is where an unsheltered lit match is easily blown out by a gust of wind or quenched by rain. Fuzees matches have been designed to continue burning in these situations
Extinction of flame	Blowing out the flame voluntarily (by blowing or shaking)	Immediate environment

Afterglow	Quenching of residual afterglow	Immediate environment. Nature of fire retardants used to suppress afterglow

Table 4.
Reactions occurring in a burning match head.[xxxv]

Solid- and liquid-phase reaction equations of potassium chlorate and oxygen-rich catalysts (MnO_2) with the main fuels (sulphur, red phosphorus and phosphorus sesquisulfide) are well known (see chemical equations below). However, different hypotheses exist about the fate of red P which is probably converted into the more reactive white form upon initial frictional heating. Potassium chlorate is inevitably decomposed (reduced) to potassium chloride in all reactions while sodium chlorate is never used due to its hygroscopic tendency. The total available oxygen in the match composition (oxygen balance) should always be in slight excess of the requirements for complete oxidation of fuels and binders. An excess oxygen content of 15% is recommended, especially for matches required in unfavourable climates. Too much chlorate on the other hand may lead to violent ignition and eventually may even prevent propagation. The formulation of match heads is still considered a trade secret by some manufacturers and many ingredients of no functional value are added as mere fillers, some of which may act as organic fuels. The amount of oxygen required must consider all of these extra components. Following ignition, the main sustaining reaction is provided by potassium chlorate reacting with sulphur (main fuel) and sulphur sesquisulfide, if present in the formulation.

$$2\ KClO_3 + 3S \rightarrow 2\ KCl + 3SO_2$$

$$4\ MnO_2 + S \rightarrow 2Mn_2O_2 + SO_2$$

$$4\ KClO_3 + (red\ P \rightarrow P_4) \rightarrow 4\ KCl + P_4O_{10} + O_2$$

$$10KClO_3 + 4P_3S_4 \rightarrow 10KCl + 3P_4O_{10} + 16SO_2$$

Combustion Temperature of a Match Head

A match head produces a small and very hot pyrotechnic flame. Pyrometric studies of burning match heads containing 41–43 weight % potassium chlorate, indicate an initial flame temperature ranging from 1,400–1,900°C (±10%) but the flame eventually cools off to about 1500°C when steady state burning of the match head is reached. Once ignited, the wooden splint burns at a much lower temperature (700–900°C) typical of hydrocarbon-fuelled diffusion flames. The flame produced by a match head is surprisingly bright and has been shown to yield about 150 lumens/second of light output, hence more than a typical candle flame.

Related to friction matches are 'electric matches' or e-matches, which are nowadays produced by the thousands and used to light firework displays remotely. They normally consist of small electrical resistors dipped into heat-sensitive pyrotechnic compositions similar to a safety-match head but much faster burning (Figure 43).

Figure 43. Electrically fired 'e-matches' used by the author to light professional firework displays in Italy.

These 'matches' require a hefty current pulse to set them off (for safety reasons) and ignite very reliably giving an almost explosive flash rather than a steady burn. Reliability and repeatability are required to ignite complex 'burst to the beat' firework displays (with music-assisted ignition).

Other Sources of Chemical Ignition

Other than matches, a number of mixtures of oxidisers and fuels in powder or liquid form can also be used to reliably make fire. Potassium permanganate and glycerine (or sugar) can be ground together to obtain a fiercely burning purple flame within seconds. Nowadays, the best survival kits also contain vials of potassium permanganate powder and glycerine as an emergency fire-starting tool. The same mixture is used by firefighters in Australia and the US to deliberately start controlled bush fires in the path of an oncoming wildfire, with a technique known as 'Delayed Aerial Ignition Deployment – DAID'. The small controlled fires starve the larger wildfire of fuel before it can reach towns and villages. Aboard a helicopter, standard ping pong balls (made of nitrocellulose) containing a small amount of potassium permanganate powder, are injected with glycerine using a hypodermic needle syringe just before launching. A number of 'primed' balls are then dropped onto the target area whereby they ignite within a few seconds of landing, setting fire to the dry vegetation. Potassium permanganate is a strong oxidiser capable of releasing enough oxygen for the complete combustion of the glycerine fuel, depending on quantities mixed together. Atmospheric oxygen is not required to cause ignition therefore this mixture functions as an improvised pyrotechnic composition. Other common oxidisers include hypochlorite, perchlorate and nitrate salts, as well as concentrated

(>70%) nitric acid. Although these compounds are not used in survival emergency kits, they may be found in chemistry laboratories, swimming pool stores and workshops, and have been known to be the cause of many unwanted fires. Any spilt residues coming into contact with fuels in a bin (i.e. paper, sawdust and organic materials in powder or liquid form) can, in some cases, cause a fire even hours after being discarded. For this reasons, fuels and oxidisers are always kept separate in modern chemistry laboratories and residues are never mixed together.

THE ADVENT OF PORTABLE LIGHTERS

During the last 200 years, smoking has been, and still is today, a popular albeit demonstrably unhealthy pastime which always requires a 'light'. Lighters were developed alongside matches mostly as a portable source of fire to light cigarettes, cigars and pipes.

Traditional lighters are portable fire-lighting devices designed to produce a flame almost instantaneously and on demand. A lighter generally yields a flame rather than an ember, therefore the mechanical 'strike-a-lights' of the seventeenth and eighteenth centuries are not classed as true lighters. However, exceptions exist with modern electric lighters. Lighters normally consist of a small reservoir (or tank) supplying a flammable liquid or gaseous fuel, and a miniaturised source of ignition. The igniter must be able to produce a tiny electrical or mechanical spark still capable of reaching temperatures high enough to reliably ignite the gas or vapour fuel. In some cases, a fine platinum wire catalyst acts as the heating element for the fuel vapour ignition process.

The first (almost) portable self-lighting device was indeed the Döbereiner lamp invented in 1823, the year that arguably marks the birth of a whole new method to make fire, at a time when flint and steel still universally prevailed. But before the invention of the synthetic flint, which represents the true apogee of lighters development technology of the twentieth century, other kinds of ingenious contraptions made their way onto the European and American markets in the form of early portable lighters. The 1890s saw a great number of designs which functioned without pyrophoric flint. These early lighters were often made of sheet brass or steel and relied on cap ignitors typical of early fire arms (amorce caps) or catalytic oxidation of methanol fuel on platinum sponge or filament. Other less common devices relied on friction-match technology in the form of 'igniting match or pellet cases' which produced a lighted match or pellet by friction, on demand. An early electric lighter, the Luminus Paris was also developed around this time which relied on a small zinc/carbon battery to make a platinum filament glow. This ignited a 'pilot' wick which in turn ignited a larger flame. Arguably the most successful of all these devices were Amorce cap lighters and, to a lesser extent, catalytic methanol lighters.

Amorce cap lighters contained a small paper strip or rotating cardboard disc onto which small dried droplets of impact-sensitive 'fulminating' pyrotechnic mixture, normally based on red phosphorus and potassium chlorate, which would flash when struck. These lighters would normally rely on a mechanism to semi-automatically open the metal lid, advance the paper roll or disc to expose a fresh (unfired) cap, and automatically strike the latter with a small steel hammer. The resulting flash would be hot enough to reliably ignite a liquid hydrocarbon fuel-impregnated wick. Fuel evaporation from the tank, which was normally packed with cotton wadding, was minimised by closing the lid after use, which would also extinguish the flame. A number of cap lighters were patented in Europe and the US in the 1890s. The best known are probably the Magic Pocket Lamp (US), Little Gem (US), Fulmen Pinel (Belgium), Doppelmond by Köhler (Germany) and other less known brands. Some of these lighters were in production for only a short time, and have now become very rare and highly collectable pieces, often fetching considerable sums at auction. The author is the very proud owner of a solid brass 1892 Köhler lighter (Figure 44), which never fails to amuse friends, neighbours and relatives when it ignites, flashing and generating a big puff of smoke.

Figure 44. A close up of a solid brass 1892 Köhler lighter, showing the cap striking hammer.

Catalytic (or platinum) lighters made a first appearance at the turn of the century and were still in use, particularly in continental Europe, well after the development of synthetic flint lighters into the early 1940s. These were normally made of a brass or steel double barrel construction, to house the fuel (methanol) and the catalyst compartments separately. Very fine platinum wires, often suspending a tiny compressed platinum sponge pellet, would be threaded between the arms of a small steel frame attached to a removable lid. Insertion of the frame into the fuel chamber containing a cylindrical wick, which was impregnated with methanol, caused the platinum filaments to glow red hot, eventually igniting the methanol vapour. This in turn would produce a small, blue and hence poorly visible flame to hover over the fuel compartment. Replacement of the lid would extinguish the flame and protect the filaments for storage. Like Amorce cap lighters, catalytic lighters were eventually confined to the museum following the introduction of the synthetic flint. But unlike Amorce cap lighters, which burnt hydrocarbon fuel, these devices produced a relatively cool flame easily extinguished by the weakest of breezes. The flame was also difficult to see in bright daylight, and the filaments were excessively delicate, often breaking off the frame upon touching, or if the lid was dropped by accident. Also the catalyst could become 'poisoned' after long periods of inactivity, requiring heating with a clean flame from another methanol lighter to burn off impurities. Despite the setbacks, these lighters played a part in the history of fire lighting and deserve a mention because of their underlying chemistry, akin to that of the Döbereiner lamp of 1823. Catalytic platinum oxidation is still used to this day in the production of portable hand warmers and relies on the intrinsic property of platinum metal, especially when divided in fine powder or filament form, to function as a catalyst for the oxidation in air of small fuel molecules (hydrogen and methanol chiefly). The metal catalyst itself is chemically unaffected by the reaction and therefore is never consumed. In this way a catalytic lighter never requires the filaments to be replaced, unless they get damaged by mishandling. The main exothermic reaction involves oxidation of methanol to give formaldehyde vapour. The heat liberated by the reaction is sufficient to make the filaments glow red hot (750°C) because of their thin diameter. The glowing filaments in turn ignite the methanol vapour/air mixture contained in the fuel compartment (and produced by the high surface area cylindrical wick) with a satisfying 'pop', thus making the ignition of this lighter fascinating to watch. The author owns rare examples of catalytic lighters dating from the 1920s to the 1940s (Figure 45).

Figure 45. Operation of a 1920s German DRP catalytic lighter. Slow extraction of the catalyst holder from the fuel compartment, causing the filaments to glow red-hot and ignite the methanol vapour.

All of them required extensive 'cleaning' of the Pt wire over an external flame before they functioned again after many decades of inactivity.

THE DEVELOPMENT OF SYNTHETIC FLINT

It wasn't until the invention of the ferrocerium alloy by the Austrian chemist Carl Auer von Welsbach in 1903, that modern lighters as we know them today could be developed. The new alloy contained iron (from the Latin, Ferrum) and the rare earth element cerium, which is very easily oxidised. The first alloys contained around 30% iron and 70% cerium but other variants were soon developed which contained lanthanum and other heavy metals to produce hotter sparks. When struck or scraped with a harder metal (like a knife blade or file), these brittle alloys flake off producing a shower of very small particles that ignite pyrophorically, producing very hot and bright 'sparks' (incandescent particles burning at approx. 3,500°C). The pyrophoric nature is due in part to the presence of cerium and in part to the small size of the ejected slivers, which require little mechanical energy to raise the temperature to incandescence. At around 500°C, chemical oxidation takes over from the frictional heat imparted, thus generating extremely bright and hot sparks. On contact with these, flammable vapours and gases ignite with relative ease. Interestingly, the name 'flint' is still affectionately but erroneously in use to this day to describe ferrocerium alloy. The term clearly dates back to the days of the strike-a-light, which employed a real natural flint to strike sparks out of steel.

The first prototypes of ferrocerium flint lighters were developed using small slabs of pyrophoric alloy which were abraded with miniaturised files to produce sparks in proximity to fuel-soaked wicks (naphtha). Eventually, slabs were replaced with small cylindrical rods of different length, which could be loaded into spring-loaded mechanisms with a small thumb-operated abrasive steel wheel (the 'thumb wheel') used to abrade the 'flint' and thus produce sparks. Because of its ruggedness, effectiveness and simplicity, the thumb-wheel ignitor is still a preferred design in most modern flint lighters to this day, especially disposable models. The first thumb-wheel lighters possibly originated during WWI when soldiers would spend long hours in trenches crafting petrol lighters out of spent brass ammunition cartridges. These lighters are still collectable today and are known as 'trench art' lighters. The years between the two wars saw a number of different brands of petrol lighters being brought to market in Europe and the US. Notable developments included the discovery, possibly by US soldiers, that a pierced brass ammunition cartridge may be used as an improvised wind shield for a flame, which eventually led to the birth of the iconic Zippo windproof petrol lighter in 1932.[xxxvi] The company is still thriving today, and offers a lifetime guarantee on each new lighter. In the early 1920s, another striker mechanism known as the 'permanent match' was developed, relying on a removable stylus metal scraper holding a wick impregnated with fuel, and a 5 cm long flint rod positioned on the side of the lighter case. On striking the flint with the scraper, the stylus wick would re-ignite from the resulting sparks. Reinsertion of the stylus in the lighter case after use would extinguish the flame and soak the scraper with fresh fuel, hence providing what appeared to be a permanent 'match' action. Striker lighters remained popular in the US until the 1950s ('Match Lite' was a famous brand), and cheap brass variants of the design are still produced in the Far East as novelty petrol lighters to this day. Petrol lighters were at their peak between 1930 and the early 1950s when the introduction of compressed butane gas as a cheap by-product of petroleum fractionation and Viton seals, for sealing fuel tanks, gradually caused a shift in the market. In contrast to petrol (naphtha) which is prone to gradual loss by evaporation (petrol lighters were rarely airtight because of the wick being present), butane gas is only released by valve action (by depressurising the tank) and also produces a much cleaner and odourless flame. Smokers immediately preferred the new technology, especially when integrated into elaborate and valuable case designs, sometime made of precious metals like silver and gold. Famous companies like Dupont, Ronson and Dunhill switched over to butane fuel and started producing beautiful and long-lasting portable and table lighters which still work fine today after 70 years.

Figure 46. Pocket and Table petrol lighters from the author's collection, dating from early 1920s to late 1940s. Both of these table lighters were produced by Ronson in England.

In the 1960's, some were made out of plastic, rather than metal, for the first time (Figure 46).

Figure 47. Ronson 'Comet' gas pocket lighter from 1968. With a robust plastic case, the chromed metal lid slides to reveal a large, semi-automatic flint wheel and two spare flints.

A number of smaller and less renowned companies produced some interesting hybrid models relying on electricity to ignite petrol-soaked wicks.

Figure 48 shows a table lighter made by Dorset Lighting Industries (DLI, UK) which was marketed as the 'flintless lighter'.

Despite its simplicity, this lighter never took off on a grand scale and working examples are rarely found today.

Figure 48. 1950s DLI 'Flintless' electrical table lighter with its outer case removed to show the battery compartment and step-up transformer. A high voltage spark is produced on removal of the stylus, which ignites a petrol-soaked wick.

Excellent reference books have been written on the subject of vintage lighters,[xxxvii] including striker lighters.[xxxviii] Ferrocerium flint ignition was, and still is, used in applications other than cigarette lighters, including relighting miners' lamps (Figure 49), camping gas lanterns (Figure 50), handheld sparking tools to ignite oxyacetylene torches for mechanic's workshops, and ferro-rod fire starters which can set leaves and magnesium shavings alight with a shower of hot sparks. This list isn't exhaustive.

Figure 49. A solid brass Eccles Type 6 'relighting' miners lamp, with close up of the flint wheel. Pulling the toggle at the side rotates the wheel, causing sparks to ignite the lamp wick. The lamp burns up to 20 hours on a tank of 'Colzaline' fuel.

Figure 50. A Campingaz C200 automatic flint ignition gas lantern. These are no longer in production and have been superseded by models fitted with piezo-electric ignitors.

Figure 51. 'Ferro-rod' emergency fire starter popular among survival enthusiasts. Some models have a magnesium block which can be shaved with a knife to produce highly flammable metal dust. Magnesium dust ignites directly from ferro-cerium sparks, generating a fiercely hot flame.

Throw It Away

The early 1970s saw the introduction of the non-refillable plastic disposable lighter, with the French company Bic leading the European market with its omnipresent oval-shaped ergonomic design.

Figure 52. Disposable Bic gas lighter.

Equipped with a large thumb wheel, an adjustable flame height control valve, a long-lasting flint rod and a sizeable butane tank, these lighters were robust, reliable and cheap enough to take over the market and compete with friction matches on an unprecedented scale. Since the beginning, sales of Bic lighters have kept soaring to six million units a year despite strong competition from the Far East, with several imitation designs now being sold worldwide. Bic now also sells a 'mini' version of the classic design, which is also disposable. Despite the huge success of disposable lighters, the sales of friction matches (SAW or safety types) have retained a niche market in many countries, due to personal preference for carrying out tasks other than smoking. Lighting fireplaces and old boilers may be easier using a long match rather than a butane lighter. Or, at least, this may be perceived to the be the case. Today, although flint lighters still dominate the market for smoking equipment, other sources of ignition incorporated into 'novelty' lighters are often preferred, especially among younger smokers. A wide range of new petrol lighters are still available for purchase due to nostalgic and other reasons, including collecting, personal preference, or just interest in the subject.

Modern-day Lighters

The 1970s saw the introduction of the piezoelectric ignitor which is in wide use at the time of writing (2020). Piezoelectricity is the property of certain crystals (including quartz) to produce strong separation of electrical charge across opposite faces when rapidly compressed. The resulting high voltage spike is harnessed to produce a small electrical arc which is hot enough to ignite butane/air mixtures. Miniaturised ignitors rely on tiny piezoelectric crystals compressed by a hand-operated, spring-loaded hammer mechanism which is often housed in a small plastic or metal case (Figure 53).

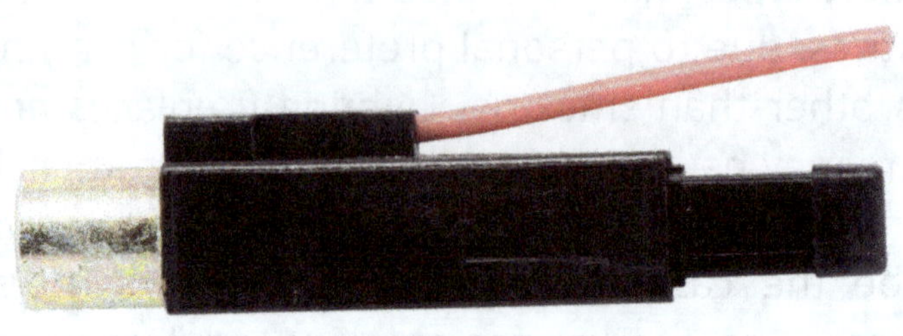

Figure 53. A new piezoelectric ignitor and same unit mounted inside a plastic refillable gas lighter. Depression of the black button causes a spring-loaded hammer to strike a small quartz crystal, causing a high voltage arc (~1kV) to jump across the gas nozzle.

These devices are also installed in other self-lighting devices, i.e. self-igniting blow torches, gas barbeques, gas boilers and camping stoves being a few examples (Figure 54).

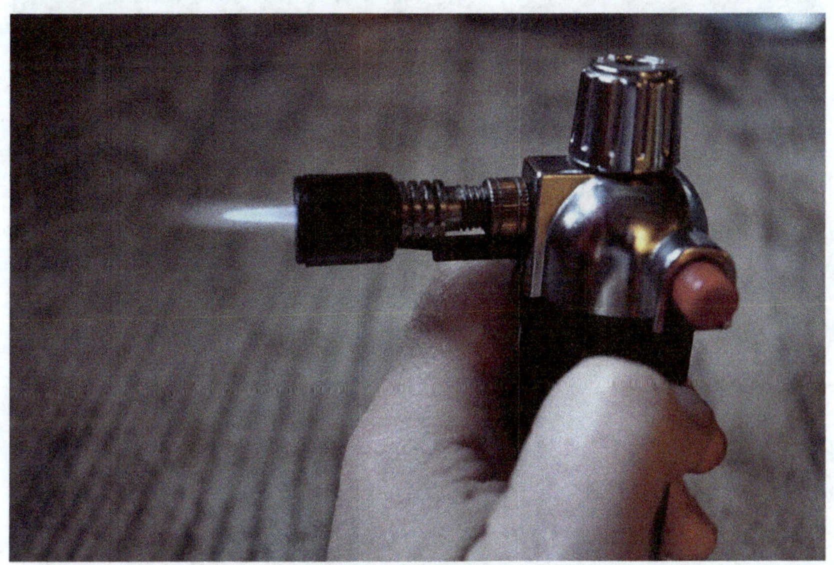

Figure 54. A small self-lighting gas torch fitted with piezoelectric ignitor.

The piezoelectric ignitor is also invariably found in novelty cigarette lighters producing other than a normal flame. Typical novelty lighters include blue 'jet' flame types, also known as 'torch lighters' which produce a very hot (1,300°C) premixed flame, sometimes of different colours (green and red) by incorporation of flame-colouring pebbles positioned above the gas nozzle (Figure 55).

Figure 55. A novelty green flame jet lighter, which also doubles as a pocket light torch.

The pebble incandesces and the thermal excitation of certain elements (boron and lithium) imparts a green or red colour to the (normally) blue flame. This is not a pyrotechnically coloured flame but the result of weak atomic excitation arising from heating up the metal salts present in the pebble. Multi-flame torch lighters also exist which produce two or three coloured flames for the amusement of their owners.

Arc or Tesla Lighters

Another very recent development in lighter technology includes 'arc' lighters, which incorporate miniaturised tesla coil circuitry to produce one or more electrical mini-streamers jumping between two close electrodes. A 'streamer' differs from an instantaneous spark in that it continues to discharge through the air until the circuit is broken (normally by releasing the ignition button). Various designs are now available (Figure 56), ranging from single-arc candle lighters featuring flexible probes to reach a tea light or other candle in a jar, and twin- or multi-arc lighters which are primarily designed for lighting cigarettes.

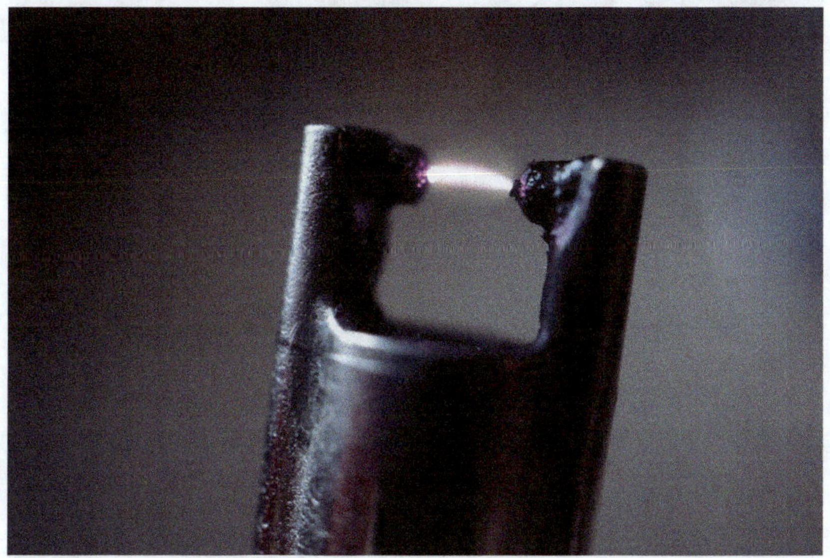

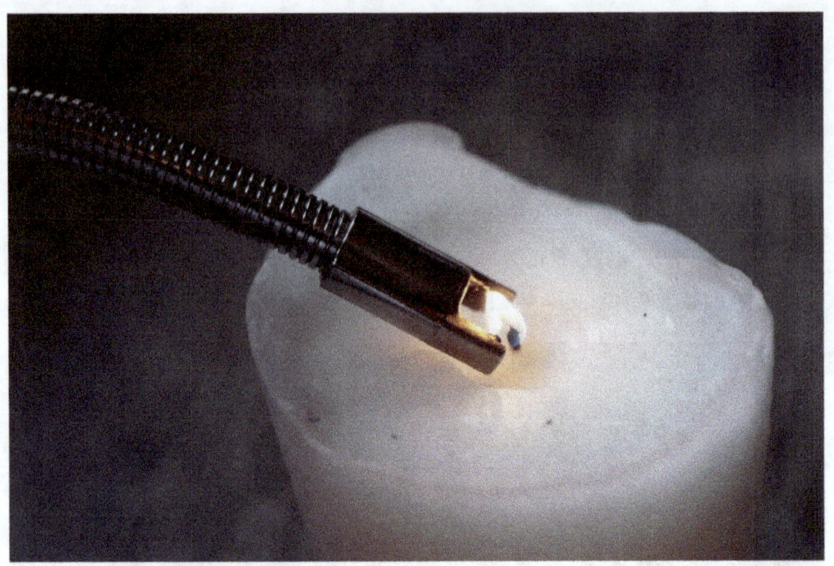

Figure 56. Single- and double- arc lighters. An elongated arm allows direct lighting of candles.

The streamers are hot (~3000°C) ionised plasma that can directly ignite paper, cardboard, wax-impregnated wicks and other material on brief contact. This technology was only made possible by recent advances in the miniaturisation of electronic components which allows the incorporation of high frequency oscillator circuitry into very small volumes, i.e. in the case of a lighter. Most of these lighters are now cheaply mass produced in the Far East and exported to the West.

Subsequent modern development includes another type of electric lighter resembling a mini-electrical hob, or mini nickel-chromium coil which glows at around 800°C from the passage of current provided by a rechargeable lithium-Ion battery. These lighters do not produce a flame and therefore are only intended for lighting cigarettes. In some hybrid models (Figure 57), both coil and arc ignition are present in the same unit, thus offering the choice to light cigarettes in two different ways.

Figure 57. A 'hybrid' Tesla lighter with arc or coil function. Both rely on the same Li-ion battery.

Coil lighters function in same way as 12V pop-up lighters found in cars, batteries may be recharged several hundred times by plugging into a USB socket for a few hours. Since USB sockets are now ubiquitously found on all laptops and desktop PCs, and are also incorporated into home power sockets, the is little risk of not being able to recharge this lighter. The obvious advantage of electric lighters is that they don't require any fuel or synthetic flint to operate.

Looking Ahead

Gas and electric lighters represent thousands of years of evolution in fire-lighting technology and humankind should be proud to have mastered and perfected the art of fire lighting for our many daily uses. We have come a very long way from the days when our early ancestors took embers back to their caves. But despite the effectiveness of all modern lighting methods, the journey is far from over. Small but powerful, battery-operated burning laser pointers already exist, and although current prototypes can only make paper smoulder at a distance of a few metres, substantially more powerful batteries will be available in the next few years. In less than a decade it may be possible to light someone's cigarette from a distance of 100 metres, hopefully without also burning a hole in their head in the process! In fact, any new methods developed in the future will have to be intrinsically safe to comply with ever more stringent legal safety requirements. Beyond that, it is possible to imagine a remote future when 'fire' will no longer be required for transport, heating, power generation and other uses. Instantaneous induction cooking hobs and electrical cars are already with us today. Smoking, fireworks and fire lighting (fireplaces and stoves) may ultimately become illegal in most modern countries in order to avoid indoor and outdoor air pollution. Fire, as we know it today, may become an accidental occurrence to avoid at all cost. And ultimately, we are still lucky to live in an era when we can still legally own a wood burning stove, have barbeques with friends and hear the roar of a petrol engine under the bonnet of our sports cars. We clearly can't stop progress and indeed must act together to minimise pollution by reducing our carbon footprint globally, but future generations may lose great life experiences based on fire which today we take for granted. Like when it's cold outside, and we have a hot drink in front of the fireplace, or when enjoying a slice of pizza baked in a traditional wood-fired oven, or when camping with friends, playing guitar around the camp fire. Long may we continue to enjoy our ancestral connection with fire for many centuries to come!

Figure 58. Relaxing in front of the fire on a chilly winter's day.

Figure 59. Pizza is best when cooked with real fire!

Figure 60. Remote wilderness camping relies on solid fire lighting and fire building skills.

ACKNOWLEDGEMENTS

The author wishes to thank ex and current work colleagues for all their support and encouragement, Mr Dale Collett of the British Bushcraft School, for his expertise and invaluable demonstrations of the hand drill method, Wing Liu for help with photographs and editing.

Also, many thanks to German archaeologist, Jurgen Weiner for many emails exchanged on the subject of early firelighting, and for providing literature and samples of Baltic flint and marcasite nodules.

He also wishes to thank the editorial staff at MTP publishing, in particular Keith Abbott and Karolina Robinson, for their patience, and Susan Hunt for proofreading with a good sense of humour.

REFERENCES

[i] J. Needham, *Science and Civilization in China*, Vol 5, Part 7. Military Technology; The gunpowder epic, 1–365.

[ii] C. Perles, *Preistoria del Fuoco*, Alle origini della storia dell'uomo, Einaudi Editori , Torino, 1983.

[iii] S.R. James, 'Hominid use of fire in the lower and middle Pleistocene.', *Current Anthropology*, 30(1), 1989, 1–26.

[iv] S. Weiner et al., 'Evidence for the Use of Fire at Zhoukoudian, China.', *Science*, 281 (5374), 251–53, 1998.

[v] K.P. Oakley, 'Evidence of Fire in South African Cave Deposits', *Nature*, 174, 261–62, 1954.

[vi] J.A.J. Gowlett, 'The discovery of fire by humans: a long and convoluted process', *Philosophical Transactions of the Royal Society B (Biological Sciences)*, The Royal Society Publishing, 2016.

[vii] P.J. Heyes et al., 'Selection and use of manganese dioxide by Neanderthals.', *Scientific Reports*, 2016.

[viii] D.A. Stopler et al, 'A Pleistocene ice core record of atmospheric oxygen concentrations, Science, 353 (6306), 1427–30.

ix A.J.L. Harris et al, 'Discovery of self-combusting volcanic sulphur flows.', *Geology*, 28(5), 2000, 415–18.

x M. A. Uman, *Lightning*, General Publishing, Toronto, 1984.

xi E. Stauffer, 'A review of the analysis of vegetable oil residues from fire sample debris samples: spontaneous ignition, vegetable oils and the forensic approach.', *Journal of Forensic Science*, 50(5), 2005, 1–10.

xii A.P. Aldushin et al, 'On the transition from smouldering to flaming.', *Combustion and Flame*, 145(3), 2006, 579–606.

xiii W.N. Watson, 'Early fire-making methods and devices, From the Stone Age until the introduction of the match.', Gibbon Brothers publishing, Washington DC, 1939.

xiv M. Christy, The Bryant and May Museum of Fire Making Appliances, Catalogue of the Exhibits, Bryant and May Ltd, 1926.

xv Glenys Crocker, *The Gunpowder Industry*, Shire publications, 2002.

xvi J. Weiner, 'Friction versus Percussion: some comments on fire-making from old Europe.', *Bulletin of Primitive Technology*, 23, 10–16, 2003.

[xvii] J. Weiner, 'Pyrite Versus Marcasite or Is everything that glitters pyrite?', *Bulletin des chercheurs de la Wallonie*, 37, 51–79, 1997.

[xviii] D. Stapert and L. Johansen, 'Making fire in the Stone Age: flint and pyrite.', *Geologie en Mijnbouw*, 78, 147–164, 1999.

[xix] K. Spindler, *The Man in the Ice: The preserved Body of a Neolithic Man Reveals the Secrets of the Stone Age*, Phoenix, London, 2001.

[xx] A. Teather and A. Chamberlain, 'Dying embers: Fire-lighting technology and mortuary practice in early bronze age Britain.' *Archaeological Journal*, 173(2), 188–205, 2016.

[xxi] A. Sorensen, W. Roebroeks, A. van Gijn, 'Fire production in the deep past? The expedient strike-a-light model.', *Journal of Archaeological Science*, 42, 476–86, 2014.

[xxii] H. Ishikawa et al., Investigations on lignins and lignification: XXVIII. 'The degradation by *Polyporus versicolor* and *Fomes fomentarius* of aromatic compounds structurally related to softwood lignin.', *Archives of biochemistry and biophysics*, 100 (1), 140–49, 1963.

[xxiii] J. Riuz-Herrera, *Fungal cell wall: Structure, Synthesis and Assembly*, CRC press, Boca Raton, 1992.

[xxiv] J. Ridgeon, *Fire by Friction – The Hand Drill Method*, Amazon Publications, 2015.

[xxv] D. Collett, 'How to make a hand drill without tools or flint.', *Bushcraft & Survival Skills Magazine*, Issue 55, April 2015, 70–5.

[xxvi] K.P. Oakley, 'Fire as a Palaeolithic Tool and Weapon.', *Proceedings of the Prehistoric Society*, 21, 36–47.

[xxvii] R. Mears, *Bushcraft – The Bestselling Guide to Surviving the Wilderness*, Hodder & Stoughton, London, 2002.

[xxviii] V. Cacciandra and A. Cesati, *Fire steels*, Umberto Allemandi & C., Turin, Italy, 1996.

[xxix] W.T. O'Dea, *Making Fire – Wood Friction, Tinder Boxes, Matches*, Her Majesty's Stationary Office, London, 1964.

[xxx] J. Caspall, 'Making fire and light in the home pre-1820.', *The Antique Collectors Club*, Woodbridge, Suffolk, 1987.

[xxxi] A.A. Mills and R. Clift, 'Reflections of the burning mirrors of Archimedes.', *European Journal of Physics*, 13(6), 268–72.

[xxxii] J. Wisniak, 'Matches, The manufacture of fire.', *Indian Journal of Chemical Technology*, 12, 2005, 369–80.

[xxxiii] W. Nicholson, *American Edition of the British Encyclopedia*, Mitchell, Ames and White, 1821.

[xxxiv] D. Thomas, *Strike-a-light*, John Walker 1781–1859, Autopress, Stockton, UK, 1990.

[xxxv] C.A. Finch and S. Ramachandran, *Matchmaking: Science, Technology and Manufacture*, Ellis Horwood Ltd, Chichester, 1983.

[xxxvi] D. Poore, Zippo: *The Great American Lighter*, Schiffer Publishing, 1997.

[xxxvii] P. Brooks, *Vintage Lighters, 100 Great Mechanisms*, Lighter Club of Britain, 2015.

[xxxviii] C.C. Sanders and W.E. Sanders Jr, *Striker Lighters*, Sea Sand Enterprises, 2003.